Solving Problems with Microscopy

Solving Problems with Microscopy

Real-life Examples in Forensic, Life and Chemical Sciences

Edited by

John A. Reffner
John Jay College of Criminal Justice
USA

Brooke W. Kammrath
University of New Haven
West Haven
CT, USA

The right of John A. Reffner and Brooke W. Kammrath to be identified as the authors of the editorial material in this work has been asserted in accordance with law.

Registered Offices
John Wiley & Sons, Inc., 111 River Street, Hoboken, NJ 07030, USA
John Wiley & Sons Ltd, The Atrium, Southern Gate, Chichester, West Sussex, PO19 8SQ, UK

For details of our global editorial offices, customer services, and more information about Wiley products visit us at www.wiley.com.

Wiley also publishes its books in a variety of electronic formats and by print-on-demand. Some content that appears in standard print versions of this book may not be available in other formats.

Library of Congress Cataloging-in-Publication Data
Names: Reffner, John A., editor. | Kammrath, Brooke W., editor. | John Wiley & Sons, publisher.
Title: Solving problems with microscopy : real-life examples in forensic, life and chemical sciences / edited by John A. Reffner, Brooke W. Kammrath.
Description: Hoboken, NJ : John Wiley, [2024] | Includes bibliographical references and index. | Summary: "Problem solving is essential to all progress. The scientific method and logical reasoning ensure that problem solving is systematic and based on factual data. One of the most significant and useful tools for solving problems is the microscope. There are many different types of microscopes, each having broad applications in a diversity of different disciplines"-- Provided by publisher.
Identifiers: LCCN 2023018048 (print) | LCCN 2023018049 (ebook) | ISBN 9781119788201 (hardback) | ISBN 9781119788218 (adobe pdf) | ISBN 9781119788225 (epub) | ISBN 9781119788232
Subjects: LCSH: Microscopy--Technique. | Microscopy--Case studies.
Classification: LCC QH207 .S65 2024 (print) | LCC QH207 (ebook) | DDC 570.28/2--dc23/eng/20230624
LC record available at https://lccn.loc.gov/2023018048
LC ebook record available at https://lccn.loc.gov/2023018049

Cover Design: Wiley
Cover Image: © Kuntalee Rangnoi/EyeEm/Getty Images

Set in 9.5/12.5pt STIXTwoText by Integra Software Services Pvt. Ltd, Pondicherry, India
Printed and bound by CPI Group (UK) Ltd, Croydon, CR0 4YY

C9781119788201_101123

Dedication

Our children, John R. Reffner, Elaine (Reffner) Teeters, Riley Kammrath, and Grayson Kammrath, and grandchildren, who inspire us to be our best selves, to make every day joyful, to think creatively, and to strive to use small (microscopical) things and thinking to make a big impact.

Our spouses (past and present), Sally E. Reffner, Phyllis Bender Reffner, and Matt Kammrath, who have provided much needed support and encouragement throughout our various scientific adventures.

Our mentors and mentees, too numerous to name, who have each left their traces on us as educators and life-long students. There is one person in the galaxy of individuals we have had the privilege of working with who shines brightest with her intelligence, integrity, creative problem solving, generosity of time and knowledge, and kindness – Dr. Pauline E. Leary – we are forever grateful for your friendship.

John A. Reffner and Brooke W. Kammrath

Contents

List of Contributors

Fran Adar
Instruments SA Inc.
Metuchen, NJ, USA

and

*HORIBA Scientific
HORIBA Instruments Incorporated
Piscataway, NJ, USA
*Current affiliation

Elena Basso
Department of Scientific Research
The Metropolitan Museum of Art
New York, NY, USA

Stephan X.M. Booerrigter
Curia Indiana
West Lafayette, IN, USA

and

*Triclinic Labs, Inc.
Lafayette, IN, USA
*Current affiliation

Richard S. Brown
MVA Scientific Consultants, Inc.
Duluth GA, USA

Sylvia Centeno
Department of Scientific Research
The Metropolitan Museum of Art
New York, NY, USA

R. Christopher Spicer
Gallagher Bassett
St. Simons Island, GA, USA

John B. Crowe
US Food and Drug Administration
Forensic Chemistry Center
Trace Examination Section
Cincinnati, Ohio, USA

Peter R. De Forest
Forensic Consultants
Ardsley, NY, USA

and

John Jay College of Criminal Justice
New York, NY, USA

Brian J. Ford
Hon FRMS Hon FLS
Cambridge, England

S. Frank Platek
US Food and Drug Administration
Office of Regulatory Affairs, Office of
Regulatory Science
Forensic Chemistry Center
Ohio, USA

Andrew Anthony Havics
pH2, LLC
Avon, IN, USA

Brooke W. Kammrath
University of New Haven
West Haven, CT, USA

and

Henry C. Lee Institute of Forensic Science
West Haven, CT, USA

Gary J. Laughlin
McCrone Research Institute
Chicago IL, USA

Pauline E. Leary
Noble
Stevensville, MD, USA

Henry C. Lee
Henry C. Lee Institute of
Forensic Science
West Haven, CT, USA

Federica Pozzi
Department of Scientific Research
The Metropolitan Museum of Art
New York, NY, USA

and

*Centro per la Conservazione ed il Restauro
dei Beni Culturali "La Venaria Reale"*
Torino, Italy
Current affiliation

Dale Purcell
Curia Indiana
West Lafayette, IN, USA

and

*Chemical Microscopy, LLC.
West Lafayette, IN, USA
Current affiliation

Nicola Ranieri
US Food and Drug Administration
Office of Regulatory Affairs, Office of
Regulatory Science
Forensic Chemistry Center
Ohio, USA

John A. Reffner
John Jay College of Criminal Justice
New York, NY, USA

Paul J. Schields
Curia Indiana
West Lafayette, IN, USA

Luis Soto
AT&T Bell Labs
NJ, USA

Mark R. Witkowski
US Food and Drug Administration
Forensic Chemistry Center
Trace Examination Section
Cincinnati, Ohio, USA

Foreword

It would take some time to think of a technique of greater utility and practical value to the physical and biological sciences than microscopy. In spite of this, its full benefits to researchers and analysts alike remain, more often than not, unrecognized and underutilized. This rather surprising failure to fully exploit the capabilities offered by the microscope and its attendant methods is due, in no small part, to those practitioners who are content to merely use microscopes but never become proficient microscopists.

Since the microscope's earliest employment by curious amateurs, such as van Leeuwenhoek, the microscope has helped mankind expand human vision ever further down in scale to reveal the most minuscule secrets of nature in much the same way that the telescope made it possible to gaze into the sky and visualize the vast expanse of the cosmos. Almost immediately following its discovery, the microscope began to reveal the true nature of the previously unseen and unimagined world that has surrounded us for millennia, by providing direct visual evidence, which no amount of philosophical argument could refute.

In the years following its invention, some microscopes were built primarily as works of art and still others were fashioned for those rich enough to purchase one merely for their own amusement.[1] However, from nearly the time of its invention, the *raison d'être* for this instrument has been to extend human vision; primarily for the advancement of knowledge and more specifically that knowledge we regard today as scientific.

It should come as no surprise, therefore, that over the passage of centuries, the capabilities of both microscopes and microscopy, have improved dramatically. Far from being restricted exclusively to advances in design, magnification, and resolution, there have been developed entirely new types of microscopes and microscopies. With them the microscopist is no longer restricted to merely observing only color and minute morphological details. The instruments available to today's microscopists permit them to not only observe and probe microscopic specimens using ordinary white, polarized, and fluorescent light but also to monitor their interaction with electrons, other electromagnetic radiation (e.g., X-rays) and even sound. Thus, these new microscopes and their accessories allow us to analyze and not just observe the most minute specimens elementally, elucidate their chemical composition and even map the distribution of discrete phases of almost any type

---

1 For instance, Samuel Pepys recorded the purchase of a microscope in his famous diary, shortly after he bought a copy of Robert Hooke's then newly printed *Micrographia*.

of microscopic feature or particle, whether of chemical, biological, anthropogenic, or even extraterrestrial origin.

Concurrent with these developments in instrumentation have been dramatic improvements in the interpretation of the results obtained by means of these instruments. Prominent among these significant new developments are those relating to the interpretation of the observations and data resulting from microscopical study. Of particular value is the availability of computerized databases (based on and developed from authentic reference collections) that include almost every imaginable substance in existence. In spite of this, not every type of substance can be identified by its elemental, chemical, or crystallographic properties. Pollen grains, wood, cellulosic and natural protein fibers such as hairs, etc. are all, for the present at least, still best identified by their characteristic microscopic morphology, which may be by enhanced by the resolution of electron microscopy and then identified by comparing the salient microscopic features to images in the authoritative atlases (some old and some new) that now exist. The availability and practical importance of such non-hardware resources helps to simply substantiate the fact that not all of the improvements to the instruments and accessories employed by the microscopist are related to hardware.

If the identification of microscopic unknowns were the sole contribution of microscopy to the solution of real life problems, it would still be of sufficient importance to justify its place as one of the premier problem solving tools of modern science and engineering. I am speaking here of real problems, not those involving merely simple decision making. As those whose profession it is to solve serious problems know all too well, there commonly occur in real-life, questions of such importance, complexity, and difficulty (and quite often also secrecy and urgency) for which solutions must be obtained, at almost any cost. The key to solving problems such as these is to be found in the realm of the specialists, who have by training, experience, and inclination prepared themselves to attack and solve such enigmas when they arise within their area of expertise. While such specialists most often hold specific titles (e.g., cryptanalyst, analytical chemist,[2] consulting engineer, medical diagnostician, etc.) they may all be considered, at least for the purposes of this essay, as *analysts*.

Successful analytical microscopists bring not only their expertise (based on years of training and practical experience) to bear on the solution of the varied problems presented to them, but in addition, develop a special, nearly unique, insight into the way the world, and everything in it, appears and operates on the microscopic scale. This rare and most unusual perspective is one of the fundamental differences between microscopists and other analytical scientists. As a result, the analytical approach that has evolved from this uncommon perspective is, as far as I am aware, a unique and rarely appreciated or fully understood advantage enjoyed by the microscopist who works as a problem solver. I am speaking here in the pedagogical sense, since in everyday practice the expert microscopist thinks in this way without ever *thinking* about it, so to speak.

The extent to which microscopes, their accessories, and resources for interpretation have been brought to perfection in our day is difficult to comprehend even for those, such

---

2 Today, the analytical chemist seems to have become a rare specialist with most analytical scientists today identifying themselves as spectrographers, chromatographers, mass spectroscopists, etc.

as I, who have worked in this field for their entire professional career. During this time (and beginning even earlier in my home laboratories as a young boy) I have sought out, consulted, and acquired an extensive collection of books on practically every aspect of microscopy that might be of assistance to my work as an analytical microscopist – both antiquarian and modern as well as practical and theoretical without regard to the language in which they are written. While I cannot claim to have or even be aware of every book on the subjects bearing on the field of applied microscopy, I can state with some authority, that I have never stopped seeking out those potentially helpful titles, of which I am still unaware, for acquisition.

This is what makes the book you now hold in your hands so special and unusual and, I will add here, important to the aspiring adept. I am unaware of another book (old or new) whose authors have undertaken the task of exploring and attempting to explain the use of the microscope as an aid in the general solution of problems – period. There are books that explain how to use the microscope in fields ranging from the brew house to barber school and to how understand the underlying causes of glass breakage. T. E. Wallis in his excellent *Analytical Microscopy* (through three editions) attempted to broadly cover a range of topics that would help to solve the types of problems that might be encountered by the public analyst in Britain in the 20th century. The *Journal of Analytical Chemistry* published, some years ago, a series of articles (*The Analytical Approach*) by analytical chemists that described the processes of logical thought and details of analysis that went into solving a variety of real-life problems in the industrial and pharmaceutical industries, art forgery detection, and forensic science among others. Books and articles such as these are certainly better than nothing, but they only tease the imagination and make the absence of a comprehensive treatment of the subject more obvious – or so one would think. This work is, therefore, long overdue and the authors are to be congratulated on conceiving and bringing to print what will become, I believe, essential reading for anyone who uses or contemplates using microscopy to help solve problems that involve materials of any sort. It has been ably written by a team of two professional microscopists with quite different but complimentary backgrounds.

I have known and admired John Reffner for most of my adult life. He is the preeminent analytical microscopist of our age and has accumulated, quite literally, a lifetime of experience in applying the microscope, in all of its manifestations, to the solution of problems, many of which would never have been solved had it not been for his clever and resourceful intervention. He began his career in the microscopy laboratory of one of the big rubber companies,[3] before being recruited by Walter McCrone. There he gained experience and sharpened his microanalytical skills, before leaving to attend graduate school. This was followed by a period in academia and afterward as the head of the microscopy laboratory of a large chemical company. Since than his reputation for solving "insoluble" problems has continued to grow and his activities have ranged from providing expert witness testimony in criminal and civil proceedings to his personal involvement in the development of the first practical, commercial infrared microspectrophotometer. There are, I believe, very few

---

3 This was in the days when most manufacturing companies had laboratories devoted to both microscopy and analytical chemistry.

applications of microscopy with which he has not at one time or another been called upon to draw from in the solution of problems of the most diverse kinds.

His co-author, Brooke Kammrath, is a professor of forensic science at the University of New Haven, where she specializes in the application of new and newly developed micro-scopical and microanalytical instrumentation directed toward the improvement in and interpretation of microscopic trace evidence. She is, compared to John, relatively new to the field, but has the great advantage of having been his student. What she lacks in experi-ence, however, she more than makes up for in energy and enthusiasm and, of course, her experience will only increase over time. It is fair to say that this book would not ever have been written if it had not been for her role in pushing it ahead. I must mention here that our first meeting was at a scientific conference where a colleague introduced her to me as an enthusiastic graduate student who had recently discovered the allure of microscopy as a practical problem-solving tool. I recall her excitement at the time, which was due to the fact that she had just purchased an elaborate polarizing microscope for her own personal use. She cheerfully relayed to me how she couldn't wait to begin to learn how to use it so she could examine any and all microscopic objects she encountered whenever she had the urge to do so. In my experience, the best microscopists would never consider life complete without their own microscope(s). Little did I know at the time that she and John would come up with the idea for a book of great importance to the field of applied microscopy that had not yet been written, but should have been, long before now.

I would like to conclude by recommending this book as an essential resource for anyone who uses microscopes or contemplates using them to solve problems. By anyone, I mean not only those in the physical and biological sciences, engineering, and medicine but also anyone who might benefit from an understanding of how the information obtained with the aid of a microscope can be put to practical use. I only wish that this book had been writ-ten when I was just beginning my career in analytical microscopy.

Skip Palenik
President and Senior Research Microscopist
Microtrace LLC

Preface

Although a plethora of books exist about the science of microscopy, most focus on descriptions of microscopical methods and instrumentation. These "how-to" instructional books are exceedingly useful for learning how to achieve a quality magnified image, but there is something missing. With this book, we are addressing the "why-to" use a microscope to solve problems. Interpreting magnified images requires a knowledgeable understanding of not only how the image was achieved, its illumination and the morphological features present, but also an awareness of how the magnified image is related to solving the problem at hand. In many instances, this requires in-depth education, training, and experience to equip a scientist with a breadth of knowledge. Whether it is a question of identification or comparison, a microscope is a sophisticated tool that requires the user to understand **how** and **why** to recognize the meaningful features or differences in a magnified image.

In this book, specific case examples demonstrate the value of using the microscope to solve problems. These cases range from criminal and civil forensic investigations to industrial, environmental, cultural heritage, pharmaceutical, and biological problem solving.

John A. Reffner and Brooke W. Kammrath

Abbreviations

AFM	atomic force microscopy
API	active pharmaceutical ingredient
ASTM	American Society for Testing and Materials
ATR	attenuated total reflection
DLI	depolarized light intensity
DNA	deoxyribonucleic acid
DSC	differential scanning calorimetry
DTA	differential thermal analysis
EDX	energy dispersive X-ray spectroscopy
ELISA	enzyme-linked immunosorbent assay
fcc	face-centered cubic
FCC	Forensic Chemistry Center (US FDA)
FIB	focused ion beam
FID	flame ionization detector
FT-IR	Fourier transform-infrared
GC	gas chromatography
GPC	gel permeation chromatography
GRS	government rubber styrene
HPLC	high pressure liquid chromatography
IA	image analysis
IR	infrared
IRRFC	infrared reflectance false color
LIB	laser induced breakdown
Mn	Number Average Molecular Weight
MS	mass spectrometry
MW	molecular weight
Mw	Weight Average Molecular Weight
NAA	nucleic acid amplification
NCE	new-chemical-entity
O-PTIR	optical photothermal infrared spectroscopy
PCA	principal component analysis

PCR   polymerase chain reaction
PLM   polarized light microscopy
PMB   polymer modified bitumen
PMRB   postmortem root banding
qPCR   quantitative polymerase chain reaction
SBS   styrene-butadiene-styrene
SEM   scanning electron microscopy
SIMS   secondary ion mass spectrometry
TEM   transmission electron microscopy
TOA   thermal optical analysis
UV-Vis   ultraviolet-visible
UCL   upper confidence limits
XRD   X-ray diffraction
XRF   X-ray fluorescence
Y-STR   Y-chromosome short tandem repeat
ZBH   zero-background holder

Introduction

Mankind's progress is historically paced by the ability to solve problems. The discovery of fire, the wheel, the steam engine, etc. were important developments in the history of humanity because they provided solutions to problems. Fire supplied warmth and led to the development of cooking food. The wheel provided ancient Mesopotamians a method for doing work at an unprecedented pace and load. The steam engine started the industrial revolution. In today's technologically advanced society, solving problems is still essential to our progress.

How do problems get solved? History tells us that problem-solving requires applying knowledge to develop tools and methods specifically to accomplish a necessary task. The requisite knowledge is gained through education, training, and experience. The scientific method provides a systematic approach for problem solving successfully. The scientific method is an organized and iterative step-by-step pathway for answering questions and solving problems. Although there are many different descriptions of those steps, these authors have outlined the following nine steps for proper application of the scientific method: observation, documentation, preservation, examination, contemplation, speculation, verification, conclusion, and communication. Table 1 details these steps of the scientific method. A critical step of the scientific method is the verification stage, which includes a feedback loop from which hypotheses and conclusions can be refined.

Defining the problem is a critical first step to developing a solution. A poorly defined problem cannot be properly solved. Observation, the first step of the scientific method, is a paramount tool for understanding all elements of the problem. Contextual information (Chapter 11) is also required for success. Although this often goes unrecognized, successful problem solvers understand the importance of stating the problem.

Solving problems requires sophisticated and logical reasoning, often in the form of inferences. Three types of inferences, or thought processes, have been delineated by scientific philosophers: deduction, induction, and abduction. Deductive reasoning (also known as "top-down logic") is the determination of a conclusion based on known rules and premises (or preconditions). Inductive reasoning (also known as "bottom-up logic") is the determination of the rule based on specific premises, results, and/or conclusions. The third and lesser known inference is abduction, where the best or most likely explanation (or precondition) is determined given the rule and the results or conclusions. For more information on logical reasoning, we refer the reader to "The Sign of Three: Dupin, Holmes, Pierce"

Solving Problems with Microscopy: Real-life Examples in Forensic, Life and Chemical Sciences, First Edition.
Edited by John A. Reffner and Brooke W. Kammrath.

Table 1 The scientific method is a process that includes these steps.

The steps of the scientific method	Activities and actions
Observation	Seeing, listening, touching, smelling, tasting
Documentation	Taking notes, photographs, drawings
Preservation	Collecting, packing, labeling, recording
Examination	Inspecting, analyzing, measuring, experimenting
Contemplation	Thinking, organizing, correlating
Speculation	Developing hypotheses, brainstorming
Verification	Validating, testing, challenging
Conclusion	Establishing relationships, interpreting data, determining significance to the problem
Communication	Reporting, publishing, presenting

edited by Umberto Eco and Thomas A. Seabok (1983), which is an excellent collection of essays on critical thinking with examples from Edgar Allan Poe, Sir Arthur Conan Doyle, and Charles Sanders Pierce. The resolution to the problems presented in this book resulted from the use of each of these forms of logical reasoning, with the specific situation determining which was implemented. For example, in the case of the Yellow Rope (Chapter 10), deductive reasoning was used. In this case, the solution to the problem or the conclusion (the two ropes came from different batches) was logically inferred from the rule (different batches of polymer have different tacticity) and the premise (the known and questioned ropes had different tacticity). In the Polio Vaccine case (Chapter 2), inductive reasoning was used to determine the rule (all containers with high residual stress fail) after the scientist was provided with the precondition (these containers have high residual stress) and the result (these containers failed). The case detailed in "A Mouse, a Soft Drink Can ... and a Felony" (Chapter 5) demonstrated the process of abductive reasoning to solve its problem. After being provided with a rule (a mouse cannot damage the outside of a soda can when inside of it during the manufacturing process) and the result (microscopic observation of damage from a mouse's teeth on the outside of the soda can), the precondition was determined (the mouse was put into the soda can after manufacturing). Deductive reasoning is the most algorithmic; therefore, it has the most guaranteed conclusions (or has the most certainty), but it is sometimes considered the least insightful. Abduction has the least certainty, but is the most insightful in that it has the potential to provide causal explanations from novel observations. Induction falls between the other two in its certainty and insightfulness. All methods of inference have value when solving problems.

One of the most significant and useful tools for solving problems is the microscope. Antoine Von Leeuwenhoek (1632–1723) is credited with being the father of microbiology due to his invention of the single lens microscope. What is often not recognized is that he developed this tool as a method for solving the problem of seeing fine threads to improve his work as a draper. Subsequently, he began looking at all kinds of materials, from bees and lice to mold and pond water. His discovery of microbes and the microstructure of materials opened the doors for numerous branches of scientific study. The microscope continues to be the symbol of science and scientific disciplines throughout the world.

Why use the microscope to solve problems? The answer lies in the first step of the scientific method: observation. Microscopes are the greatest tools for performing detailed observations about an item under investigation. In addition to imaging visual features, microscopes are measuring instruments. Microscopes have the ability to measure a huge variety of physical and optical properties. The imaging and measurement capabilities of microscopes are fundamental attributes that provide essential information which enable a microscopist to solve problems.

Microscopy is the art and science of producing, recording, and interpreting magnified images. Microscopy is both an art and a science. Art is defined as "the expression or application of human creative skill and imagination" (Oxford English Dictionary, 2022). The art of microscopy is in the skills required for successful microscopical investigations. From sample preparation to focusing the microscope to achieve the best image for the specific application, microscopy requires the art of the inquisitive mind. Science is defined as both the collection of knowledge about the physical and natural world and the method in which that information is systematically learned and studied. Microscopes are useful for both aspects of science, and are versatile tools for producing and recording magnified images that contain critical data for scientific examinations that are used for solving problems.

One of the microscope's early problems was the ability to record and communicate the image seen through the microscope to others. Although sketches were essential for this, they were laborious, required artistic abilities, and were not necessarily verifiable. The pairing of photography, both stills and motion picture, with the microscope was a major advance in the use of the microscope. This created the field of photomicrography. The ability to capture magnified images stimulated the growth of the acceptance of microscopy as it created a reviewable record that brought others into the microscope. Modern-day digital photography continues the advancement of the field of microscopy, not only making it easier to capture magnified images but also by being able to record dynamic events with video capabilities. However, the interpretation of the data demands both knowledge and a perceptive mind. Because of the failure to properly use the microscope, the incorrect or inappropriate interpretation of images remains microscopy's most pivotal problem.

The microscope has broad applications. The value of magnified images is not limited to a single discipline. The microscope is beneficial to many fields, ranging from biology, chemistry, forensic science, metallurgy, minerology, gemology, materials, and environmental science, etc. Microscopical methods are multifaceted and ubiquitous. Magnified images are useful data in all aspects of an investigation.

In addition to being a symbol of science, a microscope is a primary instrument for approaching the solution to numerous problems. The microscope is the means for taking us into the micro-world. There exists a range of complexity of microscopes, from the hand lens and children's plastic microscopes to sophisticated super-resolution electron microscopes. Their main function is to produce and record magnified images. These images are data that the scientist can use to answer questions. This will be further demonstrated in Chapter 3.

There is a vast array of different types of microscopes. One method for characterizing them is by their method of interaction with the sample of interest (Table 2). This creates

Table 2 Types of microscopy and microanalysis methods, classified by their manner of interaction with the sample of interest. Microscopical methods that can be configured with both transmitted and reflected light are designated with an asterisk (*).

Photons	Electrons	Ions	Probes
Brightfield*	Transmission(TEM)	Secondary Ion Mass Spectrometry (SIMS)	Scanning Tunneling
Darkfield*	Scanning(SEM)	Focused Ion Beam(FIB)	Atomic Force (AFM)
Stereomicroscopy*	Environmental SEM	Field Emission	AFM-Raman
Polarized Light*	Electron Microprobe	Field Ion	TERS
Dispersion Staining	Auger		AFM-IR
Phase Contrast	Scanning TEM		
Interference*	EDAX		
Modulation Contrast	EELS		
Rheinberg*	Cathodoluminescence		
Confocal*	Laser Induced Breakdown (LIB) Microspectroscopy		
Comparison*			
Microspectroscopy (UV-Vis,IR,Raman)*			
Fluorescence*			
Near-field Scanning			
X-Ray*			
Photo-electron*			
Optical-Photothermal Infrared (O-PTIR)			

four categories: photon (or light) microscopy, electron microscopy, ion microscopy, and probe microscopy. The largest and most common type of microscopy is photon microscopy. Although more commonly known as light microscopy, the term *photon microscopy* is more appropriate. Photons are the energy carrier of electromagnetic waves and are the subatomic particle that interacts with matter. Thus, the photon is used to probe a sample to create a magnified image. This is also more consistent with the terminology of the other three types of microscopies, where it is the electron, ion, or physical probe that interact with the sample to create the magnified image. When coupled with other technologies (e.g., vibrational or energy dispersive X-ray spectrometers), the capabilities of the microscope are further extended.

Every reader of this book is already aware of the value of interpreting images. We recognize our spouse, our children, our house, the street we live on, our workplace, etc. The eye forms an image and the brain interprets what we see. The human mind is exceptionally adept at recognizing similarities and differences in images. These are then used for pattern matching, identification, and interpretation. This is exemplified by the ability to identify familial relationships through a likeness in features shared between a parent and child or siblings (e.g., a break of the nose or shape of the eyes). Even in the presence of large variations, such as differing hair styles or ages, similarities are able to be interpreted to identify family members. Further, the brain's skill for differentiation by minute differences is demonstrated by the capacity to discern identical twins. These skills for image interpretation are vital for the successful use of a microscope to solve problems.

Image interpretation is a skill that requires practice, education, and training. As is true for all endeavors, there are some individuals with a natural propensity for recognizing similarities and differences in images. However, with dedication and quality education and training, skills in image analysis can be improved. For example, people can be taught to read and use a map, put jigsaw puzzles together, and navigate through mazes. The more you practice, the greater improvement you will see. In the book *Outliers*, Malcolm Gladwell proposed the 10,000-hour rule – it takes 10,000 hours of practice to become an expert in something. This equates to roughly 5 years of work (8 hours per day, 5 days per week, 50 weeks per year). Doctoral programs and apprenticeships are an excellent way to get this practice, through both education and training. On the other hand, a person can work for 40 years in a field, and have been doing it wrong that whole time. A person must practice the right thing. This is why quality education and training are vital.

For a microscope to function properly, one must have an appropriate sample. The three most important factors for good microscopical analysis are: (1) sample preparation, (2) sample preparation, and (3) sample preparation. If you do not properly prepare a sample for microscopical analysis, you will not be able to achieve the desired performance of the microscope. There are no optical tricks for compensating for poor sample preparation. The sample mounted on the microscope slide is an integral part of the optical system.

There is a union of the microscope with the observer or user that is required for good microscopical analysis. There is some fundamental knowledge that a microscope user (a.k.a., a microscopist) must have to use the microscope effectively. For example, the microscopist must focus the microscope to their eyes. This is not only to achieve a clear image with the fine and coarse focus, but when using a binocular microscope, the intraocular distance must be adjusted to fit their facial structure. This, and other necessary adjustments of the microscope and illumination system, is detailed in numerous microscopy books and websites; unfortunately they are often overlooked. A microscope is not a "plug-and-play" device; it must be customized for the microscopist. If you are using the microscope properly, you will be free from any stress or strain on your eyes. You will not get a headache or residual bright spots after using a correctly adjusted microscope. It should be like wearing glasses – the microscope should act as an extension of your eyes which enables the viewing of very small (or microscopic) objects and minute details.

There is not one "right way" to set up the microscope. Köhler (or Koehler) illumination is a good place to start, but there are reasons for making adjustments. The appropriate microscope alignment is sample and application dependent. For example, when making a relative refractive index determination, the condenser aperture should be reduced to increase contrast which enables a better observation of the Becke line of a sample. However, if you are more interested in examining the finer details at higher magnifications, the resolution of the microscope can be increased by opening the condenser aperture. There is not one way to achieve the highest quality image for every sample; the microscopist must adjust the microscope based on the desired use.

There is a need for awareness that the eye and the mind are not perfect. They can be misled. One must be careful about seeing things and making connections that do not actually exist. These may be due to cognitive bias, which has been defined as "a limitation in objective thinking that is caused by the tendency for the human brain to perceive information through a filter of personal experience and preferences" (Gillis & Bernstein, 2022). Pareidolia is a special type of bias, where one interprets an image as containing a meaningful pattern or structure that does not actually exist. This is demonstrated by the "seeing" of shapes in the clouds, such as a bunny, angel, or face (Figure 1). When using the microscope, if not properly alerted to the possibilities, one may easily fall victim to cognitive biases and pareidolia. The dangers of bias demonstrate the need for following the scientific method, validating the interpretation of images like any scientific analysis. The proper adherence to the scientific method is an excellent guard against bias.

Once the cause of a problem is discovered, an implementation strategy must be developed. In the case of the XB-70 Valkyrie Fuel Line (Chapter 11), microscopy proved essential. Microscopy was not only able to define the problem (the fuel line was failing due to the filters being blocked with debris), identify the source of the problem (two types of glass fibers: one from an exterior air filter and the second from fiberglass reinforced putty), and guide in the development of remediation and prevention efforts. In this case, and many others included in this book, the microscope proved to be essential for all aspects of solving problems.

Figure 1 A photograph of a cloud configuration appearing like the profile of a human face. Reproduced with permission from Barry Cord for Kieselstein-Cord ©Barry Cord 2021.

References

Eco, U., & Sebeok, T. A. (1983). *The Sign of Three.* Dupin, Holmes, Peirce.

Gillis, A. S., & Bernstein, C. (2022, June 22). *What is cognitive bias?* SearchEnterpriseAI. Retrieved September 5, 2022, from https://www.techtarget.com/searchenterpriseai/definition/cognitive-bias

Oxford English Dictionary. (2022). Art. In *OED Online.* Oxford University Press. Retrieved September 5, 2022, from https://www.oed.com

1

Discovery with the Light Microscope

Brian J. Ford Hon FRMS Hon FLS

Cambridge, England

We owe our understanding of the modern world to a single instrument above all others – the light microscope. These days, it is often assumed that the electron microscope is at the pinnacle of microscopy, and it seems to have eclipsed the light microscope. Indeed, there are informal reports of academics claiming that light microscopes should now be consigned to the broom-closet, since electron microscopes are so superior. The monochrome images offered by an electron microscope may provide unmatched resolution, but high resolution is not what we need for the great majority of investigations. Only the light microscope can show us the color of our specimens, and color is often crucial. And it is the light microscope alone that can reveal life as it is lived. A variable-pressure scanning electron microscope can briefly show us gray images of the movements of moribund arthropods, though only a light microscope can reveal the voluptuous twisting and turning of living cells as they pursue their complex little lives, the captivating colors of crystals under polarized light, or the selectively stained microorganisms we need to identify.

The majesty of the living cell is our current focus of attention and electron microscopes have no part to play in that. Light microscopes are among the most crucially important instruments in the world of science, and one of the few you will find in every field of investigation. Not only can they solve otherwise intractable problems, but the insights they provide influence the way we interpret the world. A trained and experienced forensic analyst can identify a particle, some strange fiber, a pollen grain, or a fragment of mineral, and solve a crime in an instant. These days, authorities rarely resort to images from light microscopes; they like to have analyses and fancy graphs to illustrate their reports, whereas microscopists recognize reality.

Microscopists are curious; we look differently at the world and are insatiably inquisitive. Indeed, this is how microscopy was born. It was on March 15, 1663, that young Robert Hooke, the 27-year-old curator of experiments at the newly formed Royal Society of London, was presented with a microscope constructed by Christopher Cock, an instrument maker from Long Acre in London (Figure 1.1). Cristopher Cock flourished in the middle 17th century. He pioneered the production of compound microscopes, the details of which appear in Robert Hooke's great work *Micrographia*, published by the Royal Society in 1665. They were provided with brass fittings and covered in polished shagreen

Solving Problems with Microscopy: Real-life Examples in Forensic, Life and Chemical Sciences, First Edition. Edited by John A. Reffner and Brooke W. Kammrath.

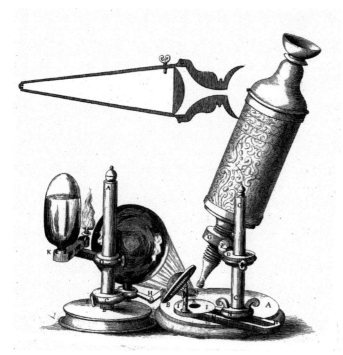

Figure 1.1   Robert Hooke's published engraving of his compound microscope.

or embossed leather. Although impressive as possessions, they gave optical results greatly inferior to those obtained with a single lens.

The Society instructed Hooke to provide weekly demonstrations to the Society's fellows. A man given to enjoying the fine things in life, he sat there with his microscope, and toying with a cork from a bottle of wine. It's strange stuff, cork. It is incredibly light, buoyant, and compressible, yet it readily springs back to its original shape. For all its porous nature, it cannot absorb fluids, so it was unrivaled for stopping up bottles of liquid. Why – since it was so light and so porous – did it not leak? Hooke decided to solve the puzzle posed by cork and wrote: "I took a sharp pen-knife and cut a thin piece of it, placing it upon a black object Plate, because it was itself a white body, and casting the light on it with a deep *plano-convex Glass*, I could exceedingly plainly perceive it to be all perforated and porous, much like a Honey-comb ..." He reported that his microscope would "presently inform me" how cork was so light, why it would never "suck and drink in water," and how was it possible to take compression more than any other substance, before it is "found to extend it self [*sic*] again into the same space." These were unique attributes, and Hooke's meticulous investigations provided the explanation. He observed that cork was composed of little boxes, or cells, "altogether fill'd with Air." Cork contained mostly empty space, and very little solid substance. It was the microscope that had revealed the truth.

1.1 Hooke, Leeuwenhoek and the Single Lens

He published his coinage of the word "cells" in his great book Micrographia, published two years later (Hooke, 1665) and the term has come down to us today. But it was wrong. To us, a cell is a living, succulent, microscopic organism and not the empty box that Hooke observed. He was identifying the empty walls inside which living cells had once existed. Far more momentous (though dismissed at the time and ignored until my revelations (Ford, 1989) more than three centuries later) was his observation of living cells in the moss *Funaria hygrometrica*. Although he wisely compared the complexity of a moss plant with that of familiar plants – like a *Sempervivum*, the houseleek – he did not mention the delicate tracery of its component cells, even though they were accurately portrayed in his fine engraving of the moss. Hooke was also the first to document a microbe, when he presented an exhibit of mildew fungi growing on old leather and recorded the details in diligent drawings. One of the paradoxes about Hooke, which scholars missed for centuries, is that you cannot observe with his microscope the fine details that he published in his engravings (Figure 1.2). I have shown that he must have used a simple – i.e. single-lensed – microscope to fill in the details, and confirmation of his methods lurks in the unnumbered pages of the Preface to Micrographia. Hooke explains how to grind and polish a tiny plano-convex lens and mount it in a metal plate. These lenses offer far higher magnification and much improved resolution though, he

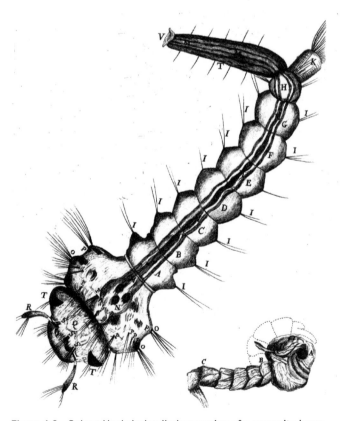

Figure 1.2   Robert Hooke's detailed engraving of a mosquito larva.

admits, they are "very troublesome to be us'd, because of their smallness, and the near-ness of the Object." (Lawson, 2016)

Robert Hooke's superbly detailed engravings were plagiarized over the centuries, and they reveal that he was an extraordinary draftsman. Here we see an aquatic larva of the mosquito Culex. Crucially, I have shown that the precise detail visible in this image cannot be seen with Hooke's microscope. Clearly, he used a single-lensed (simple) microscope – a design for which he published in his Preface – to observe fine structures which he incorporated into the final engraving.

It was this method of making a magnifier that caught the attention of the Dutch draper Thonis Leeuwenhoek, whom we know as Antony (and who added a "van" to his name in 1686, to give himself greater respectability). Although these simple microscopes were problematic, they were cheap and easy to make at home if you were a dedicated enthusiast, and Leeuwenhoek became single-minded in his quest for optical perfection. He had visited London in 1666, when he came across Hooke's great book. It inspired him to take up microscopy, and he was soon making his own little microscopes at home, based on Hooke's design (Figure 1.3).

Figure 1.3   Optical and SEM images of a newly discovered Leeuwenhoek microscope.

Details of the handmade original Leeuwenhoek microscope can be seen with this uniquely detailed study compiled from over 130 separate images obtained with a light microscope (Figure 1.3, left). For the first time, I have imaged the same instrument through a Hitachi S-3400N scanning electron microscope at the Cavendish laboratory, Cambridge University, and we can compare the lifelike image obtained with light microscopes (Figure 1.3, left) with the higher resolution of the scanning instrument (Figure 1.3, right).

When Robert Hooke wrote of Pulex irritans, the flea, he did so from personal knowledge (Figure 1.4). The large engraving published in Micrographia, measuring some 43 cm (17 in) long, was often removed from the book and framed for public view. It is extraordinarily detailed – yet, as this micrograph taken with a Hooke-type microscope demonstrates, the features that Hooke depicts are not visible with his microscope. For the fine details he resorted to the use of a single-lens, simple microscope.

To fully appreciate the detailed structure of a head-louse (Figure 1.5), Pediculus capitis, we need to use a simple, single-lensed microscope. In this micrograph, a lens ground by the late Horace Dall – a British optical specialist who ground lenses of the type that

Figure 1.4 Flea through Hooke-type microscope.

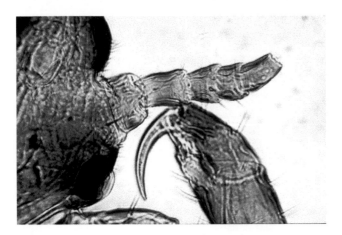

Figure 1.5 Head-Louse through single lens.

Leeuwenhoek used in his research – has been employed to show the greatly improved view that a single lens can provide. The resolution of these lenses is within a factor of four of the achromatic lenses that were to supersede them.

Leeuwenhoek began by faithfully reprising observations described by Hooke in *Micrographia*. Leeuwenhoek is widely disregarded by much present-day science teaching, though his personal inquisitiveness gave us the first glimpse of cell nuclei and sperma-tozoa, the discovery of bacteria and protozoa, and he single-handedly launched the science of microbiology. Whereas Hooke concentrated on the study of familiar objects – a magni-fied razor, a piece of moldy leather, fossilized wood, moss, seeds, and small insects from gnats to fleas – Leeuwenhoek was preoccupied by what the naked eye could not perceive. Hooke, you might say, was a macroscopist [*sic*]; it was Leeuwenhoek who was the first true microscopist.

Between them, these two irascible, independent, single-minded, and diligent inves-tigators laid down the science of investigation using light microscopes to solve prob-lems. Hooke's perplexity over the lightness and porosity of cork – combined with its extraordinary ability to resist penetration by fluids, while springing back to shape after being compressed – made him realize that only a microscope could offer the answer. Leeuwenhoek's curiosity about mucilaginous matter floating in lake-water led him to peer at it under a lens and bequeathed to us the first description of microorganisms in history.

1.2 Single-lens Microscopes come of Age

When Hooke and Leeuwenhoek died, their science died with them. There were no students to follow, and no devotees who would carry on their endeavors. When Carl Linnaeus set out to name and categorize every known form of life from 1735, he used a microscope (I have been to examine it in Sweden) yet he largely ignored the microscopical world and grouped microbes as "Microcosmus" (Ford, 2009). He didn't mention bacteria. A few philosophers recorded magnified images and made useful discoveries, including Marcello Malpighi whose observations of kidney anatomy (the Malpighian corpuscle and the Malpighian tubules) are familiar to present-day students. Jan Swammerdam used careful microdissection to elucidate the structure of insects, and in 1686 a microscope by Joseph Campani of Bologna shows in a woodcut portraying the first recorded use of a microscope in medical investigation (Ford, 2009). A handheld microscope is in use, with an assistant directing candlelight upon the area under investigation.

Giuseppe Campani was a Roman philosopher of the 17th century whose screw-barrel microscope (Figure 1.6, left, enlarged) was the first to be portrayed in medical diagnosis. This engraving of his microscope in use was published in the Acta Eruditorum published in Leipzig in 1686 by Christoph Günther (Mencke, 1686). The investigator is shown using the microscope to scrutinize the details of a wound in a patient's leg, while an assistant holds a concave mirror that reflects candlelight onto the site.

Other writers mentioned microscopy, but their claimed originality was disproved by a comparison between Hooke's and Leeuwenhoek's published illustrations and those in later books (Ford, 2010). Indeed, when Hooke published pictures of snowflakes, he was actually plagiarizing them from an earlier work by Bartholin, and when Leeuwenhoek sent to the Royal Society his reprise of Hooke's demonstrations of cork, elder-pith, bovine optic nerve and the quill of a feather, he did not acknowledge his predecessor's example. Persistent plagiarism is still prominent in today's Royal Society yet has its roots in that body's earliest days (Ford, 2020).

Figure 1.6 First use of microscope in medical investgation (Pictorial Press Ltd/Alamy Images).

When the Swiss naturalist Abraham Trembley came across *Hydra* in the 1730s, he launched into a series of diligent investigations showing how this tiny multicellular animal could be cut, spliced and grafted. Trembley was the first to carry out such extraordinary experiments yet did so without knowing of Leeuwenhoek's descriptions of *Hydra* published in 1677 (Dobell, 1932). His skill at micromanipulation using a microscope, and his remarkable prescience in undertaking far-sighted investigations, have made Trembley the father of experimental biology (Lenhoff *et al.*, 1986).

Simple microscopes of the kind Hooke designed (and which Leeuwenhoek used all his life) were handheld, though Trembley's investigations of the 1740s necessitated having the lens mounted on a stand, leaving both hands free to manipulate the specimen. An articulated arm was one answer to the problem; a fixed lens with a stage and mirror was even better. This idea was taken up in England by John Ellis, who had an instrument-maker named John Cuff produce a portable microscope that could be packed in a small box. It was popular for examining small plant specimens (and so gained the name "botanical microscope") and equally useful for observing *Hydra* and other pond organisms (instruments known as "aquatic microscopes"). The terms are interchangeable. By the early 1820s, in Regency England, these became popular possessions of a growing scientific class and their lenses provided remarkably clear images. One of the investigators who realized their potential was the Scottish surgeon Robert Brown, who named the cell nucleus in 1831 and painstakingly

Figure 1.7 Single-lensed microscope from the 1820s by Bancks and Son.

dissected the ovules of plants to study the process of fertilization. He also documented the ceaseless movement of microscopic particles, due to molecular bombardment, that we still refer to as Brownian motion. His battered and bruised microscope survived the centuries, dismissed by all as crude and incapable of serious microscopical use. Its restoration proved to be a revelation, and the images it generates compare favorably with those from present-day instruments. Single lenses are universally described as producing chromatic, rainbow-hued images, though the limitations have proved to be less pronounced in practice.

Single-lensed microscopes for research were manufactured from brass (Figure 1.7), and supplied with a range of lenses, by Robert Bancks and his son of London for luminaries including Robert Brown, the originator of Brownian motion, who named the cell nucleus in 1828; Charles Darwin, who took his on his voyage aboard the Beagle in 1835; Sir William Hooker, the first director of the Royal Botanical gardens at Kew, England; and George Bentham, the premier systematic botanist of the 19th century.

During his research on plant tissues in 1831, botanist and surgeon Robert Brown examined the upper surface of an orchid leaf under a simple microscope (Figure 1.8). He clearly discerned a small ovoid body within each cell, and named it "an areola, or nucleus." His term "nucleus" has come down to modern science. Others had observed the cell nucleus before (it was first recorded by Leeuwenhoek in the erythrocytes of fish in 1682) though no previous observer had commented upon it, or offered a name.

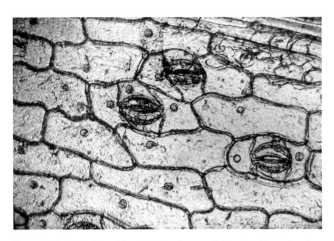

Figure 1.8   Orchid epidermis showing nuclei, under Brown's microscope of 1831.

1.3   Light Microscopes in the Modern Age

The smallest object a Leeuwenhoek lens can resolve measures 0.7 μm, whereas the resolution limit of a corrected microscope lens is >0.2 μm, and it was the introduction of achromatic lenses in 1758, and apochromats in 1763, which first allowed microscopists to approach this limit. It took decades before the advantage was accepted by microscopists (Darwin took a simple microscope on his voyage aboard the Beagle in 1831) but those improved lenses would open a new world to investigators. In 1858, a German anatomist, Joseph von Gerlach, experimentally infused brain tissue in a solution of the red dye carmine, and noted that the tissue when sectioned, showed clearly stained nuclei and he became the first microscopist to recognize the value of differential staining. The biological and medical sciences were not the only disciplines to benefit; in 1863 Henry Clifton Sorby explored metallurgical microscopy and realized the significance of carbon in creating steel from iron. His microscopical investigations led directly to the invention of the Bessemer Converter which underpinned the development of modern steelmaking.

The notion of an insatiably curious investigator, who sees the significance of traces that others cannot perceive, brings to mind Sherlock Holmes. The capacity of this fictitious detective for deriving complex truths from simple evidence is well-known, though he was never depicted using a microscope. The closest we come is in *A Study in Scarlet* (Doyle, 1888) where Arthur Conan Doyle writes of Holmes taking out "a large round magnifying glass" to search for evidence. This image has become associated with Holmes ever since, though a "large round" magnifier has low magnification. The capacity of a lens to magnify is inversely proportional to its surface curvature, and hence its radius, so a small lens can magnify more than a large glass.

Although Holmes's diagnostic skills are fanciful, they are mirrored in the real world of microscopy. The pioneer of forensic microscopy was Edward O. Heinrich of the University of California at Berkeley. In 1923 he was brought in to investigate evidence after a train robbery. He considered statements from witnesses, and examined a pair of dirty overalls recovered from the scene. Grease reported from the fabric proved to be resin from fir trees found in the Northwest, confirmed by wood fragments and pine needles in the right-hand pocket. He also found traces of mustache wax, nail clippings, and cigarette butts ... by the time he'd finished, Heinrich led the police to their suspect after informing them that he was a fastidious, left-handed Caucasian lumberjack some 5 ft 10 in tall with light brown hair and a new coat. A receipt hidden in a pencil pocket even gave a family address. This was the dawn of what we now know as CSI, a subject which has since gained an enormous public following. The annual survey by the Eurodata TV Worldwide organization in 2012 said that "CSI: Crime Scene Investigation" was the most-watched show in the world (Ford, 2015).

Today there is unremitting pressure to ignore light microscopy in a move toward digital diagnosis. It cannot work as well. Nobody can diagnose asbestos fibers in a building – or in a patient – as rapidly and reliably as a microscopist with a wise eye. In an era where tuberculosis is rapidly spreading, there are pressures to use nucleic acid amplification (NAA) tests to spot the causative organism. The commercially available Xpert MTB/RIF tests disposable cartridges of sputum that are analyzed in the GeneXpert Instrument System and can give a result within two hours. These machines cost tens of thousands of dollars and the cheapest cartridges cost over $10. The light microscope allows a smear to be examined at minimal cost with results in a matter of minutes. *Aspergillus* is regularly identified using an enzyme-linked immunosorbent assay (ELISA) test though most laboratories around the world lack the equipment. Polymerase chain reaction (PCR) tests are also available, though are not reliable. Yet an experienced microscopist can recognize *Aspergillus* spores in seconds, without even needing to stain the specimen.

Microscopists identify pollen grains and can use the characteristic surface profile to diagnose the origins of a specimen (Figure 1.9). They variously possess spikes, protuberances, and surface vesicles. These are from the conifer *Pinus*, and the pollen from those conifers have a unique structure, because attached to each haploid cell are two air-sacs in the manner of water-wings. These serve to provide buoyancy which allows each pollen grain to catch the drafts of air which facilitate their air distribution.

High magnification is not always what we need. Living pond protozoa are among the most enticing organisms to study under a light microscope, as Leeuwenhoek discovered in 1767. The use of low power – typically, less than 10× – can frequently provide a sense of structure that otherwise eludes us.

The sagittal section of a full-term rat embryo has been stained with hematoxylin and eosin (Figure 1.10). It is 15 mm long and is viewed here under low power in a study comprising four separate micrographs. Details, from the anterior cerebellum to the genitals and posterior caudal structures, yet including the kidney and suprarenal, allow us to gain a revealing view of the internal anatomy of the entire embryo. Low-power microscopy is often ignored by researchers yet can provide unique insights.

Pollen grains? We know them: there is no need for some costly analysis. Diatom frustules? They can be characteristic of a given environment and we know how to handle those. Mineral dust, or particulates? They can be easily and quickly characterized. Leaf fragments? Leave them to us. Animal hairs? No problem whatever. Paper fibers? Scraps of soil? Sand? Sawdust? For so many of these, there are costly, tedious, and time-consuming analytical techniques using sophisticated equipment. Yet there is also the knowing eye of a microscopist. Nobody needs to collect genetic data and analyze the genes to know your

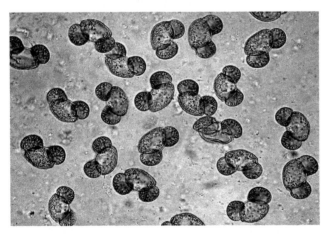

Figure 1.9 Characteristic pollen grains of *Pinus* under a modern light microscope.

Figure 1.10 The anatomy of a rat embryo under the low-power light microscope.

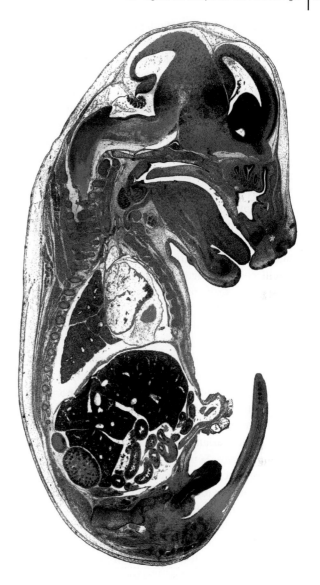

grandmother when you can recognize her in a crowd. You don't need to order PCR print-outs to distinguish an apple from an orange – you know by looking.

Not a week goes by without my using some aspect of microscopy to throw light on a problem, or to elucidate some mechanism that would otherwise elude us. Yet our modern world is hell-bent on a race to digitization and that curious concept of "artificial intelligence" which confers wondrous propensities upon costly gadgets while denying us the freedom to know what we know. I daresay there are apps that can probe our cooking and flash up on a screen when that egg is cooked – but a trained chef just knows. You can go to the laboratory and use a machine to amplify the genes in a plant and plot out its genetic profile, which could eventually be automatically linked to a botanical database that gave you the species.

Or a gardener will just tell you.

And so it is with the light microscope. Learn to use it. Study how to recognize what you see. Cultivate inquisitiveness and objectivity and be prepared to shock other people. Those blebs on the fine finish of your car? Microbes caused them. Why did that engine fail? Bacteria in the fuel. What caused those fish to act strangely? Flickering protozoa in the blood. Why did those rodents lose their fear of predators? Protozoa taking control of their brains. Have I been conned by paying a huge some for this jar of Manuka honey? Certainly, if it isn't rich with Manuka pollen grains. Can you be sure this banknote is a forgery? The printing is perfect! But its paper is made with Asian fibers, which we never use. What's this powder? Are these fibers genuine? Do I make sperm? Is this asbestos? Are these faked? Which is the right fabric? Is that genuine starch? What's in my urine? Have I got leukemia?

Ask a microscopist. They'll tell you. It won't be complicated, and it won't take long.

Could apps replace this skill? If you cannot find a light microscopist, and need some indication of what you've found, then apps will one day find the way. They do not yet exist, but within a few years they certainly will. We are told that they work through "artificial intelligence" but that's the wrong word. They aren't intelligent. What they offer is digitized automation so, if you wish to perform a routine assessment – a regular blood count, say, or a routine check for particulate contaminants – then an app will save time and tedium. But they won't eliminate the need for experienced microscopists. They alone will know where to search for what you need, how to collect the sample, how to prepare and mount the specimen, and will tell you what's there.

Apps are destined to lessen the drudgery of routine, but they aren't people. BloomOptix is the first for microscopy and identifies algae using data from over 200 microbial blooms. It is said to be 94% accurate. I have a state-of-the-art plant identifier on my phone which assures me that a pomegranate tree was a myrtle plant. I took a picture of a lei from Hawaii, an artificial garland of silk flowers, which the app assured me was the Christmas cactus *Schlumbergera*. Some years ago, a team of geneticists were puzzled by what seemed to be a miniature crossword puzzle appearing in a chromosome preparation. I recognized it as a fragment of a diatom (Ford, 1986). Diagnostic microscopy is always a challenge, and the results often pose the most unexpected juxtaposition.

The next decade will inevitably bring more automation, and an increased demand for graphs and digital data. Yet behind it all lies the microscopist. The light microscope will always entertain, illuminate, confound, and exasperate; yet – no matter how much the apps assist us – it will still be the keen eye, and that wise mind, which alone will solve the greatest diagnostic problems we face in the future.

About the Author

Brian J Ford is world-renowned as a leading microscopist, and the author of hundreds of research papers (and many books) on microscopes and microscopic life. He is also a popular television broadcaster and an international lecturer. His research has covered topics ranging from blood coagulation to paleontology, and from food science to the study of intelligence in living cells. Professor Ford has extensively researched early microscopy.

He discovered that Leeuwenhoek's original specimens still existed, was the first person to examine them through an original lens, and has recently identified two previously unknown Leeuwenhoek microscopes. He also restored Robert Brown's microscope to use and took the first photographs through it. Professor Ford has connections with several universities and is based in Cambridgeshire, England.

References

Dobell, C. (1932). *Antony van Leeuwenhoek and his "little animals." Being some account of the father of protozoology and bacteriology and his multifarious discoveries in these disciplines.* John Bale, Sons & Danielsson, Ltd.

Doyle, A. C. (1888). *A study in scarlet.* Ward Lock & Co.

Ford, B. J. (1986). Mystery solved. *Nature, 323,* 675. https://doi.org/10.1038/323675a0.

Ford, B. J. (1989). Antony van Leeuwenhoek—Microscopist and visionary scientist. *Journal of Biological Education, 24*(4), 293–299. https://doi.org/10.1080/00219266.1989.9655084.

Ford, B. J. (2009). The Microscope of Linnaeus and his blind spot. *The Microscope, 57*(2), 65–72.

Ford, B. J. (2010). The cheat and the microscope: Plagiarism over the centuries. *The Microscope, 58*(1), 21–32.

Ford B. J. (2015). Forensic science, peering down a blind alley. *The Microscope, 63*(2), 77–88.

Ford, B.J. (2020). Science? What science? *The Microscope, 68*(1),33–45.

Hooke, R. (1665). *Micrographia: or some physiological descriptions of minute bodies made by magnifying glasses, with observations and inquiries thereupon.* Martyn & Allestry.

Lawson, I. (2016). Crafting the microworld: how Robert Hooke constructed knowledge about small things. *Notes and Records: The Royal Society Journal of the History of Science, 70*(1), 23-44. https://doi.org/10.1098/rsnr.2015.0057.

Lenhoff, S. G., Lenhoff, H. M., & Trembley, A. (1986). *Hydra and the birth of experimental biology, 1744: Abraham Trembley's Mémoires concerning the polyps.* Boxwood Press.

Mencke, O. (1686). *Acta Eruditorum.* Christoph Günther.

2

When Problem Solving, Exercise the Scientific Method at Every Step

John A. Reffner, Ph.D.[1] and Brooke W. Kammrath, Ph.D.[2,3]

[1] John Jay College of Criminal Justice, New York, NY, USA
[2] University of New Haven, West Haven, CT, USA
[3] Henry C. Lee Institute of Forensic Science, West Haven, CT, USA

Introduction

The scientific method is presented in the Introduction of this book. Adhering to the steps of the scientific method provides a valuable and systematic framework for solving problems. Throughout the process of problem solving, it is important that all inquiries be directed by the science. The scientific method is critical in all aspects of an investigation, from the first stages of recognition and informed sample selection through to the final interpretation and resulting decision making and communication.

Five cases from a range of disciplines are presented in this chapter which highlight the value of science, and specifically the scientific method, throughout the process of scientific discovery.

Solving Problems with Microscopy: Real-life Examples in Forensic, Life and Chemical Sciences, First Edition.
Edited by John A. Reffner and Brooke W. Kammrath.
© 2024 John Wiley & Sons Ltd. Published 2024 by John Wiley & Sons Ltd.

2.1 The Buttonier Case

On September 26, 1974, Irving and Rhoda Pasternak were murdered in their home in Waterbury, Connecticut. The Pasternaks were an affluent family with strong ties to their community. Mr. Pasternak's body was found on the kitchen floor, with 29 stab wounds and cuts on his hands indicating he had fought with the intruder. Mrs. Pasternak was killed upstairs in her bedroom, the victim of 24 stab wounds. A blood-covered hunting knife was found nearby.

At the scene, numerous items of evidence were recognized, documented, preserved, and collected. Important evidence included bloody footwear outsole patterns that were identified as Cat's Paws soles and heels. Cat's Paws were popular after-market sole and heel replacements for shoes with a distinct logo on the heel (Figure 2.1). The footwear patterns began around the kitchen table and then continued directly up the stairs. This indicated that the perpetrator knew his way around the house. When searching the scene for trace evidence, Officer James McDonald discovered a button and two small plastic pieces on the shag carpet in the bedroom of the victim. These would later prove to be critical evidence in this case.

Murray Gold, the former son-in-law of the Pasternaks, was immediately identified as a suspect due to past history. Mr. Gold had recently approached the Pasternaks demanding more money. According to the Pasternak's daughter, this request was denied and angered Mr. Gold. Additionally, a neighbor had seen a "big blue" car near their home 3 days prior to the murder, and due to a string of break-ins, she had her brother write down the license plate information. New York license plate 833-QED was registered to Murray Gold, placing him in the neighborhood at that time. During an investigation of Murray Gold's New York City apartment, the investigators found several pairs of shoes with Cat's Paw heels, replacement buttons, and a Buttonier kit.

Figure 2.1 Cat's Paw replacement heels (exemplars, not evidence photographs).

A forensic comparison of the footwear outsole patterns from the scene to the recovered shoes with Cat's Paw heels was then attempted. In order to preserve the footwear outsole patterns at the scene, after proper documentation with photography, Officer James McDonald removed the floor boards which contained the impressions from the Pasternak's home. This forethought enabled forensic scientists to conduct a high-quality comparison to the suspect's shoes. Although a class match was made from the size and pattern, there were no individualizing features found. Furthermore, all of the recovered shoes were excluded, based on inconsistent individualizing marks on the outsoles. The Cat's Paw outsole patterns in blood combined with the same type of shoes at Murray Gold's home were useful circumstantial evidence for the prosecution, but they could not directly associate him with the crime.

As previously mentioned, a dark colored button was recovered from the shag carpet of the victim's bedroom. At the home of Murray Gold, a card which was sold with five buttons on it was found. One button was missing from the card, leaving four buttons for comparison. The questioned button from the scene was consistent in color, shape, and size with the four buttons from the card. Additionally, when the button supplier was contacted, the questioned button was identified as having been produced by the same manufacturer. This too was informative circumstantial evidence.

At the home of Murray Gold, a Buttonier kit was recovered. Buttonier kits are used to reattach a button to a garment, and consist of a tool for insertion and strip of several T-bars attached via posts (Figure 2.2). The T-bar is broken or cut from its post by the insertion tool, and the T-bar is passed through the button and garment which attaches the two together. In the Buttonier kit, there were strips that contained T-bars and posts where T-bars had been removed. The two small plastic pieces recovered from the murder scene were similar to the T-bars in the Buttonier kit.

Upon stereomicroscopic examination, Officer James McDonald observed that one of the two plastic pieces from the scene contained the remnants of its attachment to a post (less than a millimeter in diameter). He proceeded to use a stereomicroscope to compare the broken edge of the T-bar to those from the posts recovered at Murray Gold's home. He made a physical fit association between one of the posts and the questioned T-bar from the

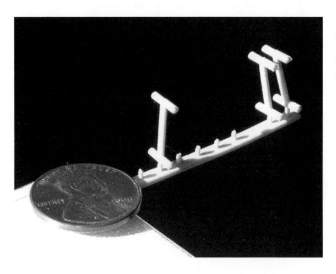

Figure 2.2   A photograph of a Buttonier kit strip with three attached T-bars and five posts with the T-bar removed. A penny is included for scale.

scene. At the first trial, Officer McDonald testified for the prosecution to these results and conclusion. However, a defense witness, Herb MacDonald, testified that he was unable to reach the same conclusion. He made an inconclusive determination based on his stereomicroscope observations of the T-bar and posts. On cross examination, Herb MacDonald was asked what he would need in order to make a positive association, to which he replied that he would use a scanning electron microscope.

The first trial resulted in a mistrial due to a hung jury in 1976, so the prosecution sought a second trial for 1977. This time, they found a consulting scientist from the University of Connecticut Institute of Materials Science, Dr. John A. Reffner, to perform a scanning electron microscopical examination of the Buttonier evidence.

Upon receiving the evidence items, the first step was to complete macroscopic and stereomicroscopical examinations of the questioned T-bar and the known posts from the Buttonier kit recovered from the home of Murray Gold. Although it was known that there was testimony of an association, it was not revealed which post an association had been made to in the prior examination. This initial examination showed that there were two types or T-bar removal from the post: cutting and tearing. These produced distinctly different morphologies on the fracture surfaces, which allowed for elimination of several posts from consideration due to the questioned T-bar having been torn from its post. Stereomicroscopy enabled one post to be identified for potential individualization with the T-bar.

Prior to performing scanning electron microscopy, the State's Attorney had to gain permission from the court to allow the consultant to gold-platinum coat the evidence items for proper imaging. This was the first case in Connecticut where scanning electron microscopy was used, and this necessary process of metal coating had to be explained due to its potential alteration of the evidence. Permission from the court was given, and scanning electron microscopical analysis followed. Figures 2.3 and 2.4 show scanning electron microscopy photomicrographs of the questioned T-bar and post from Murray Gold's Buttonier kit. Figure 2.3 shows low magnification scanning electron microscopy images of the fracture surface on the post (left) and on the T-bar (right). When increasing the magnification (Figure 2.4), similarities between the fracture surfaces are observed however they are mirror images of each other. To enable a direct comparison of the fracture surfaces, a reverse image of the T-bar was made photographically by flipping the negative (Figure 2.5).

Figure 2.3 Scanning Electron Photomicrograph of the fracture surfaces of the Buttonier post (left) and the questioned T-Bar (right).

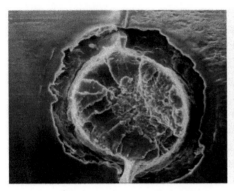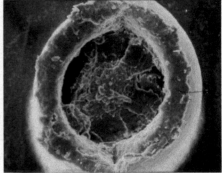

Figure 2.4 Scanning Electron Photomicrograph, at higher magnification than Figure 2.3, of the fracture surfaces of the questioned T-Bar (left) and Buttonier Post (right).

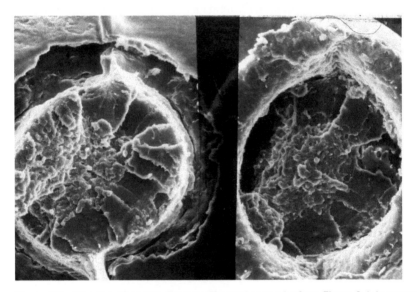

Figure 2.5 The same Scanning Electron Photomicrographs from Figure 2.4; however, the image of the fracture surface of the questioned T-Bar (left) was reverse printed to enable a direct comparison with the image of the post (right).

Interpreting scanning electron microscope images requires an understanding of how images are creating using a scanning electron beam. The scanning electron beam causes electrons to be excited and scattered. These electrons are detected and the brightness of the image is determined by the geometry of the surface. Flat surfaces appear black while sharp edges appear bright. Thus an image of a depression is darker while an elevation is brighter. Another consideration when interpreting these images is that the material of the T-bar and post are themselves elastic polyurethane polymers. When tearing the T-bar from the post, the elastic material stretches and deforms during the breaking process. Fracture surfaces are different for elastic versus brittle breakages, which is easily visualized with a scanning electron microscope. Minor distortions to the fracture edges are due to the random nature of elastic fractures and relaxation after breaking.

In the second trial, when presenting the results of the comparison of the fracture match to the court, the consulting scientist included an exhibit of the scanning electron microscope photomicrographs. Although there are some minor differences in the fracture surfaces, those can be explained by the elastic fracture deformation. More importantly there were 18 identified points of commonality of the two fracture surfaces, which were the basis of the opinion that these two pieces were at one time a single piece. This was pivotal evidence for the prosecution, and was widely reported on by the news media. This also added weight to the circumstantial physical evidence of the button and Cat's Paw outsole footwear patterns in blood at the scene.

In the field of forensic science, it is often difficult to get the court to accept new technologies. For this case, the rules of evidence were determined by the Frye Standard (Frye v. United States, 293 F. 1013 (D.C. Cir. 1923)) which is based on general acceptance by the relevant scientific community. The first commercially available scanning electron microscope, the "Stereoscan," was manufactured by the Cambridge Scientific Instrument Company and sold to Dupont in 1965. This same model was the one used in this case, which was the first one in the state of Connecticut. The use of the scanning electron microscope in this case showed how this new technology could enhance the analysis of trace evidence.

In November of 1976, Murray Gold was convicted of two counts of murder, and sentenced to two concurrent 25-year-to-life terms. In 1980, the conviction was overturned. He was retried two more times, with the third trial resulting in a mistrial and the fourth a guilty verdict. Additional details of this case can be found on the internet, and in two *New York Times* articles published in April 4, 1985, and July 25, 1986.

2.2   The Leaky Polio Virus Dispettes

Lederle Laboratories, the pharmaceutical division of American Cyanamid, was the sole supplier of the live Sabin oral polio virus vaccine in the early 1980s. Given the severity of the polio virus, production of the vaccine was of critical importance. A problem with the manufacturing of the vaccine could potentially produce dire consequences which could result in a resurgence of this crippling and deadly disease in the United States.

The Sabin oral polio virus vaccine was distributed in small plastic dispettes which were thermally sealed after filling (Figure 2.6). The seal had to be free of leakage to prevent oxygen from entering the contents, which would alter the virality of the drug. Due to the importance of this drug, it was essential to check the sealing of all of the vials prior to distribution. This testing of the seal of the virus was vital, and involved a sophisticated but simple process. A small amount of indicator, likely phenolphalein, was added to the solution prior to sealing. Then the filled dispettes were placed in a carbon dioxide pressurized chamber, and equilibrated overnight. If the seal failed, the contents of the vial would change color due to the acidification of the solution. Those vials that turned pink would be discarded. This system worked well for an extended period of time with a minimum number of rejected dispettes.

Suddenly, and without explanation, the failure rate of these dispettes increased to an unacceptable level. As a result, Lederle Laboratories would have to stop production and distribution of the Sabin oral polio virus vaccine. The dire consequences of this are important to recognize given the societal need for this vaccine. The supplier of the dispettes was consulted, and he indicated that there was no difference in the composition or process of their manufacturing.

To identify the cause of these failures, various analytical chemistry techniques (i.e., testing for changes in viscosity, additives, etc.) were used to check the chemical composition of the dispettes. These yielded no useful results due to no analytical differences

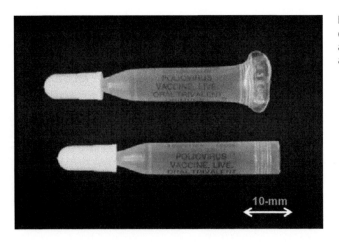

Figure 2.6 Polio virus dispettes before (bottom) and after (top) sealing, taken with a stereomicroscope.

between the samples of good and bad dispettes, where the good ones were from an earlier batch with no malfunctions and the bad ones came from new batches with a high failure rate. As a last resort, the dispettes were sent to the microscopy laboratory at American Cyanamid. In accordance with the scientific method, the first thing that the microscopist did was to look at the samples. When viewed using the stereomicroscope, the dispettes all appeared indistinguishable in terms of their geometry, morphology, color, and opacity. Next, the samples were examined using a polarized light microscope (PLM), which had both transmitted and vertical (or epi) illumination. Meaningful differences were observed in the crossed polarized light images of the good and bad dispettes (Figure 2.7). The bad dispettes displayed higher retardation than the good ones, which indicated that they had increased residual strain. A hypothesis was then developed: the increased strain in the polymer indicates that the molding process of these dispettes was altered.

To validate the observation that was made with the PLM, thermal optical analysis (TOA) was performed (also known as depolarized light intensity or DLI). TOA is an extraordinarily useful, but underutilized technique, which provides a photometric recording of the change in retardation as a material is heated (Miller, 1970; Reffner, 1975). The TOA method and instrumentation is described by Reffner (1975, p. 4) in his PhD dissertation:

> Depolarized light intensity [DLI] measurements are comparable to thermal analytical methods such as DTA [differential thermal analysis] and DSC [differential scanning calorimetry]. Changes in crystal structure or melting result in changes in optical anisotropy which in turn result in changes in the intensity of polarized

Figure 2.7 Polio virus dispettes prior to sealing, examined using crossed polarized light, (A) with a good thermal seal and (B) a bad thermal seal. The good thermal seal shows lower-order retardation colors.

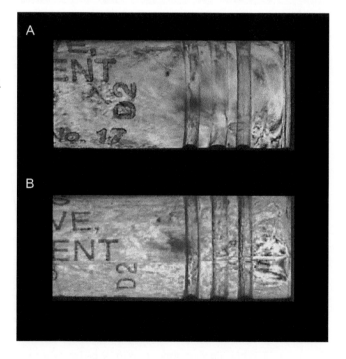

light transmission. The basic instrumentation for DLI consists of a controlled light source, a polarizer, the sample, and a second polarizer (set with its polarizing direction at 90° to the first polarizer) and a photo detector.

The TOA instrumentation (Figure 2.8) used to analyze the dispettes used monochromatic light. This enabled the quantification of the stress as it was relaxed by heating, going from high to low levels of retardation. High relative intensity values are attained when there is constructive interference, and reaches a minimum with destructive interference. The retardation is measured by the number of peaks encountered during the heating cycle. The results of the TOA of the dispettes are shown in Figure 2.9. The defective dispettes showed

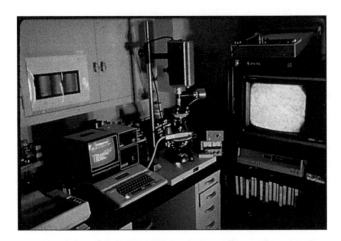

Figure 2.8 Thermal Optical Analysis Instrumentation, American Cyanamid (1983).

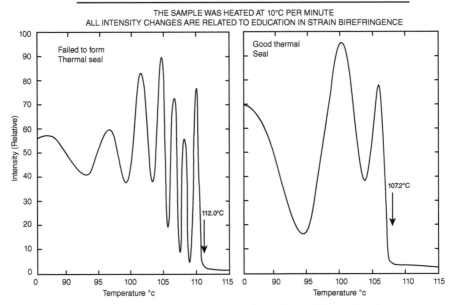

Figure 2.9 TOA of Polio virus dispettes with (A) bad and (B) good thermal seals.

seven orders of retardation (corresponding to the seven peaks), while those from an earlier batch with a good thermal seal had only three orders of retardation. This confirmed the hypothesis that the dispettes with the bad thermal seal had a higher level of residual stress.

When the data was presented to the dispettes supplier, they reviewed their records and indicated that although there was no change to the starting material, they had changed their process. In an attempt to speed up manufacturing, the supplier had increased the pressure and lowered the temperature during molding. These changes had the unforeseen consequence of affecting the thermal seal, and was corrected by returning to the original manufacturing conditions. This high priority problem was resolved quickly once the microscopist was consulted.

References

Miller, G. W. (1970). Thermal analysis of polymers. In J. F. Johnson & R. S. Porter (Eds.), *VI. Thermal Depolarization Analysis (TDA). Analytical Chem. Vol II* (p. 397). Plenum Press.
Reffner, J. A. (1975). Thermal optical analysis and microstructural characterization of polymers. *PhD Dissertation*. The University of Connecticut.

2.3 The Green River Killer

In the summer of 1982, four women were found dead in and on the banks of the Green River in Seattle, Washington (Figures 2.10 and 2.11). Over the next two decades, numerous young female murder victims were discovered in the Seattle area. Most of these victims were sex workers or teenage runaways.

The Green River Task Force was formed in August of 1982 to coordinate the investigation of these serial killings. By the early 1990s, the Green River Killer was credited with between 60 and 104 murders of young prostitutes in the Seattle area.

In 1983, Gary Ridgeway was identified as a potential suspect due to being seen with one of the victims prior to her murder. Gary Ridgeway was a paint detailer at the Kenworth

Figure 2.10 The Green River, Seattle, Washington. (Image provided by Skip Palenik.)

Figure 2.11 A photograph of one of the victims of the Green River Killer. (Image provided by Skip Palenik.)

truck factory in Seattle, with a history of frequenting prostitutes. In 1984, Gary Ridgeway passed a polygraph test and did not fit the FBI's criminal profile, hence he was not arrested.

In November of 2001, advances in DNA technology enabled the state forensic laboratory to associate Gary Ridgeway with semen found in four of the victims. Although this lead to his arrest for four of the murders, he initially pled not guilty. His explanation for the DNA results was that he had sex with these women for money, but did not kill them. This could not be refuted.

The Green River Task Force then contacted Skip Palenik at Microtrace, a private forensic science laboratory in Elgin, Illinois (Palenik, 2007a & b). Microtrace scientists were asked to undertake examinations of microscopic traces in this case, specifically paint evidence that had been collected from the scene of numerous victims on the banks of the Green River. Known samples of paint from Gary Ridgeway's vehicles and home were also analyzed and compared. After 6 months of analysis, Palenik recounts that he and his team were unable to make any associations from a few thousand individual comparisons of the paint evidence. It is important to understand that this inability to make an association demonstrates the tremendous variety of paint types that exists in the population, thus showing the value of paint evidence as meaningful trace evidence.

Palenik next recommended to the Green River Task Force that he should do a microscopic examination of the evidence, and look at the particles embedded in the clothing of victims and a pair of coveralls collected from Gary Ridgeway when he was initially identified as a suspect (Figures 2.12–2.14). The task force immediately agreed to allow Microtrace scientists to examine the microscopic traces recovered from this clothing evidence.

Upon microscopic examination, numerous small spray paint spheres of various colors were found in the debris of six of the victims and Gary Ridgeway's coveralls (Figures 2.15 and 2.16). Spray paint spheres are formed when paint is aerosolized and does not reach the target surface. Although most of the paint goes on the target, the aerosolized paint spheres settle randomly on nearby surfaces, including the painter's clothing. These paint spheres can also then be transferred to other items by physical contact.

In order to determine the composition of the spray paint spheres, Palenik took advantage of modern infrared and Raman microspectroscopy technologies. The infrared spectral analysis, Figure 2.17, identified the evidence spray paint spheres recovered from the clothing of six victims and the suspect as Imron (Figure 2.18). Imron is a high end specialty paint produced by DuPont which is not commercially available to the general public. It also had a chemically distinct formulation which enabled it to be differentiated from other paints produced at the time. Raman microspectroscopy was used to identify the pigments in the spray paint.

When the Green River Task Force was informed of this evidence, they contacted DuPont to find out where in the Seattle area they had sold Imron paint in the early 1980s. DuPont revealed that there were two companies that they supplied Imron to at that time, one of which was the Kenworth truck factory. Recalling that Gary Ridgeway was the paint detailer at Kenworth, this evidence became extremely valuable.

(A)

(B)

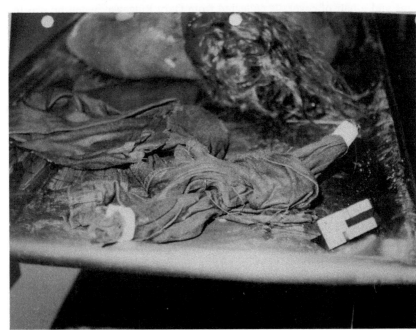

Figure 2.12 Crime scene (A) and autopsy (B) photographs of two of the victims of the Green River Killer, and their recovered clothing. (Images provided by Skip Palenik).

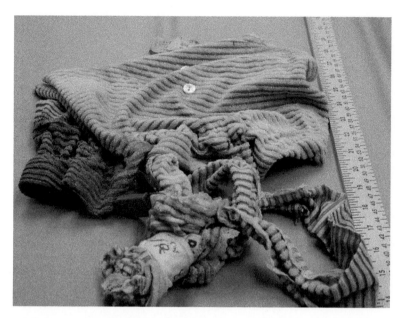

Figure 2.13 Evidence photograph of one of the Green River Killer victim's clothing. (Image provided by Skip Palenik).

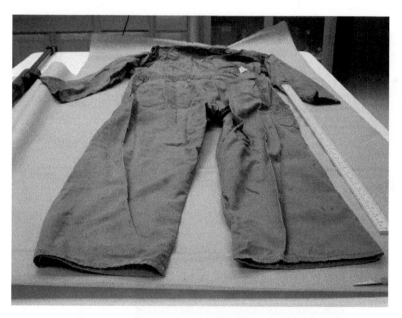

Figure 2.14 Evidence photograph of Gary Ridgeway's clothing, seized in the 1980s. (Image provided by Skip Palenik).

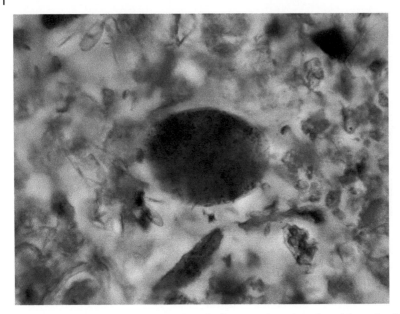

Figure 2.15 Brightfield photomicrograph of spray paint sphere found from the debris collected from the clothing of Gary Ridgeway. (Image provided by Skip Palenik.)

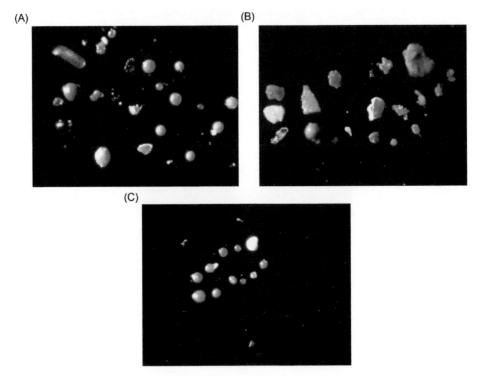

Figure 2.16 Stereomicroscope photomicrographs (A–C) of spray paint spheres found from the debris collected from the clothing of several of the victims. (Images provided by Skip Palenik).

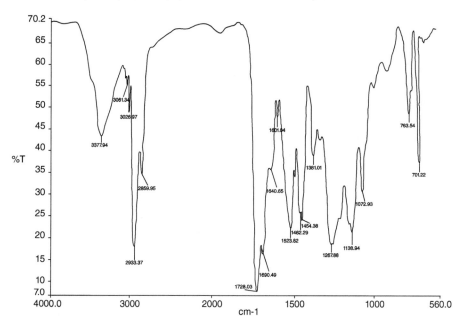

(A)

MICROTRACE

Description: dupont imron-"polyurethane enamel" source scott ryland

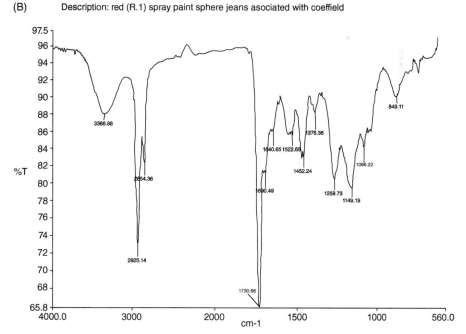

MICROTRACE

(B)   Description: red (R.1) spray paint sphere asociated with coeffield

Figure 2.17   Infrared spectrum of DuPont Imron "polyurethane enamel" (A) and of one of the spray paint spheres recovered from the clothing of one of the early Green River Killer victims (B). (Images provided by Skip Palenik).

Figure 2.18 Imron Paint Cans. (Image provided by Skip Palenik).

The significance of this evidence cannot be understated. Two of the six victims whose clothing had the microscopic spray paint spheres were also associated to Gary Ridgeway based on the DNA evidence. Thus now there were eight victims connected to Gary Ridgeway by trace and/or DNA evidence. Furthermore, the variety of spray paint colors found on the clothing of the six victims and the limited distribution of this type of paint made this circumstantial evidence stronger. It is less likely that these women had access to this specialty spray paint in multiple colors on their own, as opposed to having been in contact with Gary Ridgeway who used this paint daily in his work.

When confronted with the results of the microscopic trace evidence analysis contained in Palenik's report, Gary Ridgeway took a plea deal and confessed to 48 murders to avoid the death penalty. In November of 2003, he was convicted of 49 murders and sentenced to life in prison. Later, Gary Ridgeway admitted to strangling between 75 and 80 women, and is the second most prolific serial killer based on confirmed victims in United States history.

A valuable lesson from this case is the potential for trace evidence to provide investigative leads for criminal cases. In 1985, Skip Palenik was teaching forensic microscopy in Washington State for Dr. Walter C. McCrone. At that time, George Ishi, the director of the Washington State forensic laboratory, discussed with Skip the Green River murders, and indicated that when a suspect was identified that they would hire him to examine the evidence to aid in the prosecution. However, had the investigators had the insight to examine the clothing evidence back in 1985, Skip would have performed the same microscopical and microspectroscopic analysis that he did in 2001. Thus the spray paint spheres would have been recognized and their manufacturing source would have been identified. This would have led to DuPont and subsequently the Kenworth truck factory in 1985, thus moving Gary Ridgeway the top of the suspect list and bringing him to justice at that time. This case demonstrates that the use of trace evidence for providing investigative information in a criminal case was, and continues to be, grossly underutilized.

References

Palenik, S. (2007a, August). The contributions of chemical microscopy to the solution of the green river murders. *Presentation delivered at the Trace Evidence Symposium.*

Palenik, S. (2007b, November). The contributions of chemical microscopy to the solution of the green river murders. *Presentation delivered at the Eastern Analytical Symposium.*

2.4 The Unfortunate Failure of the Dragline Excavator

In the 1960s, a coal miner in southern Indiana obtained a contract with a local utility company to supply coal from his private mine. This was dependent on timely delivery of coal, and failure to meet the agreed upon deadlines could void the contract. Initially, he was able to meet his performance schedule using a dragline excavator. A dragline excavator is a large mobile crane equipped with a bucket that is swung out, dropped to the ground, and pulled back to collect the desired material from the ground. The filled bucket is then maneuvered to empty the collected material into a pile or container. A dragline excavator is commonly used in strip mining coal. One day, the dragline excavator suffered a bearing failure that prevented its use. The coal miner quickly contacted a supplier in Sandusky, Ohio, purchased a replacement bearing, and installed it to get the dragline back in operation. On completing the repair, the dragline excavator was operational for approximately four hours before failing again. The coal miner's decision was that he made a mistake by not replacing the shaft as well as the bearing, thus he ordered replacements for both. He drove to Sandusky again to obtain a new shaft and bearings to try to expedite repairs. The new shaft and bearings were installed, and again the dragline was put back in operation. Unfortunately, with the new shaft and bearings, failure reoccurred after about four hours of mining coal. The miner determined that the only other possible explanation was that the Sinclair lubricant was causing the failure. So, he purchased another new shaft and bearings, as well as a new lubricant from a different supplier. After replacing the failed parts with the new shaft and bearing, the Sinclair lubricant was flushed out and new lubricant was pumped through the system. Upon restarting the dragline excavator, it performed without failure. Sadly, due to all the delays, the utility company terminated his contract which effectively put him out of business.

The coal miner sought an expert opinion to confirm his suspicion that the lubricant was responsible for the dragline excavator failure, and he hired a metallurgist in an adjoining state to investigate the case. The metallurgist performed an incomplete visual examination of the bearings and falsely concluded that it was a malfunction of the lubricant. As a result, the coal miner consulted an attorney and filed a civil litigation suit against Sinclair Oil Corporation.

Sinclair Oil Corporation hired McCrone Associates in Chicago, Illinois, to examine the evidence consisting of oil-covered bearings and the shaft. The Technical Director, who was also the principal microscopist, took responsibility for this case investigation. The initial examination of the bearings consisted of analysis using a stereomicroscope. The bearings were composed of a copper alloy. The bearing surface was not immediately visual because of the dark lubricant on the surface. In order to see the damage to the bearing, the grease was removed using a solvent. The surface of the bearings showed striations (scratches) that ran around the circumference of the bearing. This circumferential damage indicated an abrasive action occurred while the shaft rotated around the bearing. If it had been a lubricant failure, these striations would not have been present because it would be a smooth surface impact between the shaft and bearing. Upon further examination, microscopic silver-colored metallic balls (~1 mm in diameter) were seen embedded in the bearing surface. There were also pitting in the surface from where these balls had impacted the bearing but not remained. As a result of seeing these impacted microscopic metal balls and the empty pits, the opaque lubricant was carefully extracted with solvent and microscopically examined. Using the stereomicroscope, the small metallic balls were seen in the extract from the lubricant.

A metallurgist was consulted to understand the origin of these microscopic metal balls. His response was immediate, as he readily recognized them to be shot peening balls. Shot peening is a metal-working process, similar to sandblasting, used to strengthen a metal surface. Small metal balls, called shot, are used to bombard a metal producing a plastic deformation of the surface. In this case, after visiting the shaft manufacturing facility, it was confirmed that shot peening was used in the final processing of the shaft. This process created a residue of microscopic shot throughout the facility and covering its manufactured products.

It was immediately clear that the cause for the series of second and third failures of the dragline excavator was the contamination of the shaft with shot peening metal residue which caused the bearing to fail. Thus, Sinclair Oil Corporation was not at fault because the lubricant did not cause the failure.

The civil case against Sinclair Oil Corporation proceeded to trial. The plaintiff presented his case first, which included testimony of the initial metallurgist. Upon cross examination, the defendant's attorney, in an attempt to intimidate the witness, stated "Wait till John's balls get in this case!" After all testimony was given, the jury ruled in favor of the defense which dismissed Sinclair Oil Corporation of any financial responsibility for the coal miner's loss.

Unfortunately, because the miner's metallurgist failed to make a proper initial examination, thus missing the critical evidence of the shot peening metal balls, the miner sued the wrong company. Further, the statute of limitations for filing a suit against the manufacturer of the shaft had expired, thus preventing the miner from recouping appropriate monetary damages from them. Clearly the coal miner attempted to do all of the right things to save his business, but was misled by an incompetent expert. Further, the coal miner failed to be warned about the potential for damage to the dragline excavator due to residues from the shaft's manufacturing process.

When initially hearing the coal miner's story, it seems probable that the lubricant was a potential cause for the malfunction. However, failure to adequately test this hypothesis by the initial metallurgist magnified the problems. The cause of the failure was only revealed when the scientist properly used the scientific method and examined the physical evidence with a microscope. The delay in obtaining the true reason of failure interfered with the harmed party receiving justice.

2.5 The Bodega Burglary

In the spring of 2017, a consulting criminalist was contacted via email by a prosecuting attorney regarding a burglary case in Connecticut, USA. The email from the prosecutor stated:

> An armed guy breaks into [a] convenience store late at night, breaks the window, enters and steals money, cigarettes etc ... His clothes tear and they find some fabric caught in the glass door. ... He is stopped and [ ] a pat down reveals a weapon and cigarettes, lottery tickets from the store. The [police officers] note[d] in their report that the fabric at the scene appears similar to the defendant's clothes. The defense attorney is dealing with a younger client and [his] dim parents. He has asked if a comparison can be done between the defendant's known clothes and the sample from the scene to assist in making [the] kid understand the futility of a trial.
>
> The issue is the State Lab "doesn't do fabric comparison anymore." They referred us to the FBI. The FBI will only do it "in major cases" for us. Therefore, I was wondering if this might be something you could assist us with.

After getting court approval to work on the case, the evidence was turned over to the criminalist for forensic examination. The questioned item that was recovered from the scene consisted of a black-colored yarn adhering to a glass fragment, in a heat-sealed plastic bag (Figure 2.19). Two known items of clothing from the suspect were provided for fabric comparison, and included black pants and a black North Face hooded jacket. It should be noted that both known clothing items were delivered to the criminalist in sealed paper bags, with the following note on each: "ITEM REPACKAGED IN PAPER BAG ON [XX/XX/XX] BY [ ]. ITEM ORIGINALLY SEALED IN PLASTIC HEAT SEAL PACKAGING AND MOLD/MILDEW FORMED ON ITEM. I SUSPECT THAT THE ITEM WAS WET WHEN INITIALLY SEALED IN PLASTIC." The clothing items were covered in mold, making the examination both unpleasant and more difficult than necessary had they been properly preserved when collected from the suspect.

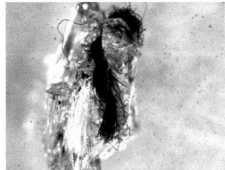

Figure 2.19   Stereomicroscope photomicrographs of questioned fibers on broken glass.

Initially, the criminalist performed macroscopic and microscopic examination of the questioned evidence item. The microscopic examination consisted a stereomicroscopic inspection to determine the construction of the yarns and general fiber characteristics. Brightfield and PLM for fiber identification and ultraviolet-visible (UV-Vis) microspectroscopy for fiber color/dye analysis was then completed. The questioned yarn was composed of black cotton fibers with a z-twisted construction.

The known clothing items were then examined. The jacket was immediately eliminated from being a potential source of questioned fibers due to it being composed of a synthetic polymer. However, the black pants consisted of microscopical and spectroscopical similar black cotton fibers with a z-twist yarn construction (Figures 2.20 and 2.21). Accordingly,

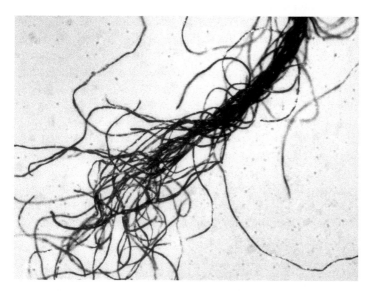

Figure 2.20 Stereomicroscope photomicrographs of known fibers from the defendant's pants.

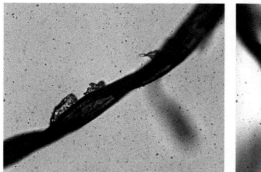

Figure 2.21 Brightfield photomicrographs of questioned fibers from the glass fragment (left) and the known fibers from defendant's pants (right).

the questioned yarn/fibers were consistent with originating from the suspect's pants or another source composed of fibers with the same microscopical and spectroscopical characteristics.

The significance of a fiber association depends on prevalence of the specific fiber. For example, white and blue cotton fibers have limited significance because of their commonness, while a bright-colored nylon fiber with unusual microscopic features would have substantial value for an association. Specific to this case, black cotton is fairly common. The questioned yarn and those comprising the suspect's pants had the same construction, which added to the value of the association, but this still had limited significance because of its ubiquity.

The criminalist went further than the initial charge for a fabric comparison, and examined the known clothing items for damage and other traces. The black pants from the suspect showed an area of relevant damage on the rear of the pants near the waistband, which consisted of five nearly parallel lines of angular scratches. Some of these penetrated the fabric while others were superficial. These lines of damage had characteristics that are observed in fabrics that have been in forceful contact with a sharp object, where either the fabric or the object is in motion. The identity of the sharp object that caused the damage was unknown, but could include a number of items such the edges of broken glass, the tip of a blade, the edge of a pointed nail, to name only a few.

When the black pants were examined with a stereomicroscope, there appeared to be glass fragments in the damaged areas (Figures 2.22 and 2.23). The hardness and morphology of a few particles were tested, and found to be consistent with glass fragments. Using crossed polarized light, the particles were slightly anisotropic possibly due to strain birefringence from the breaking event. The presence of glass on a garment of clothing has real significance. Published research has demonstrated that it is not common for individuals to have glass fragments on their clothing. Consequently, the identification of glass on

Figure 2.22 Stereomicroscope photomicrograph of the textile damage to the suspect's pants, with a small glass fragment in the lower right quadrant.

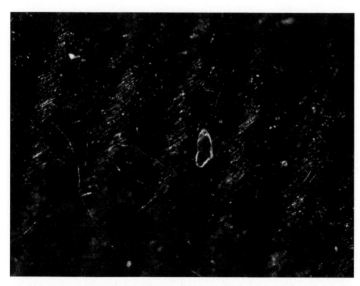

Figure 2.23 Stereomicroscope photomicrograph of a small glass fragment on the black pants.

an item of clothing indicates that the wearer of the garment either broke glass or was in close proximity to a glass-breaking event. Numerous glass fragments are easily transferred when glass is broken, and when a window is broken glass fragments are transferred both in the forward and rearward direction. Clothing items retain glass fragments well, and particles fall off over time and at different rates, depending on a number of factors including the size of the fragments, the texture of the clothing, the activity of the person, etc.

The combination of the recent damage to the pants and the presence of trace glass particles was consistent with the damage being caused by contact with the edges of a broken glass surface. Ultimately, this placed the suspect at the crime scene.

The prosecuting attorney was then called to report the initial findings and to request permission to continue with a glass examination. Because a piece of glass from the window was recovered from the crime scene, a glass comparison could be conducted which could further strengthen the case via multiple associations. The attorney was pleased with the oral report of the examination and requested a written report. However, he felt that further work on the glass was not necessary at that time. Although the criminalist would have liked to have continued the examination to confirm or refute an association between the questioned glass on the pants and known glass from the window, unfortunately this is not the way the justice system works. The prosecution defines what is presented to the court, and not the expert. Continuing with a glass examination without the consent of the attorney would have been an ethical violation.

When the teenage suspect and his "dim parents," who had initially wanted the case to go to trial, were presented with the report from the consulting criminalist, they accepted the offered plea deal. This was deemed an appropriate outcome for this armed burglary by the prosecution, the defense, and the judge who accepted the plea.

3

Images Are Real Data, Too

Introduction

When you look through the microscope, you see images. Images contain a tremendous amount of information, as expressed in the trite saying "a picture is worth a thousand words." These images can be recorded and documented, which enables the image to be shared and reviewed. It is common practice in many fields of science for images to be properly interpreted to solve problems and answer questions. For example, cancer diagnosis is often made from the interpretation of cell types in a biopsied tissue using microscopy. Different types of illnesses, most notably bacterial and parasitic diseases, are identified by microscopic imaging. Malaria is identified by imaging parasites in blood serum while Legionnaire's disease is identified by visualizing its characteristic bacteria. Images are real data, too.

Making a conclusion from an image is often considered subjective, because a person is doing the interpretation from something they see based on their knowledge. Subjective is defined by the *Merriam-Webster Dictionary (n.d.)* as "characteristic of or belonging to reality as perceived rather than as independent of mind." This is often viewed as having a negative connotation, which is inappropriate and incorrect. Using images to solve problems is dependent on the mind, thus making it subjective, but that does not lessen its value. Although science strives for objectivity, relying on the perception of an experienced person is necessary and should not be underestimated or misunderstood.

Images provide information that may be missed if only quantitative data is used. One example of this can be seen in the area of statistical analyses. Statistics are used to summarize information about a dataset, but the calculated numbers themselves do not present the whole picture. That is why graphical representations of data are critical for exploratory data analysis. Figures 3.1 and 3.2 show a collection of graphs that all have the same summary statistics (x and y means, x and y standard deviations, and Pearson correlation coefficient). One would make a mistake to only rely on the statistics when evaluating these datasets, as there is useful information contained in the graphical image which enables their differentiation. These meaningful differences are not captured in the statistics but instead are contained in the image.

Solving Problems with Microscopy: Real-life Examples in Forensic, Life and Chemical Sciences, First Edition.
Edited by John A. Reffner and Brooke W. Kammrath.
© 2024 John Wiley & Sons Ltd. Published 2024 by John Wiley & Sons Ltd.

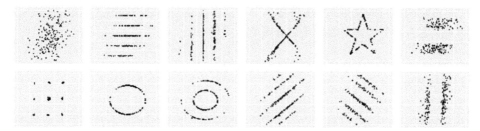

Figure 3.1 A collection of 12 datasets that are different in appearance but each has the same summary statistics: mean ($\bar{x}$ = 54.02, $\bar{y}$ = 48,09), standard deviation (sd_x = 14.52, sd_y = 24.79), and Pearson's correlation coefficient (r = +0.32). Matejka et al., 2017/Association for Computing Machinery.

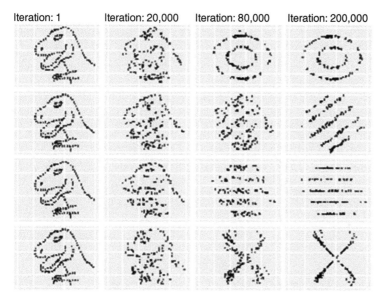

Figure 3.2 The "dinosaurus" dataset, originally produced by Alberto Cairo, which demonstrates the importance of visualizing. Although this dataset has "normal" statistics, the plot results in an image of a dinosaur. The summary statistics for the "dinosaurus" datatset are: mean ($\bar{x}$ = 54.26, $\bar{y}$ = 47.83), standard deviation (sd_x = 16.76, sd_y = 26.93), and Pearson's correlation coefficient (r = −0.06). Matejka et al., 2017/Association for Computing Machinery.

Interpreting images requires knowledge gained through quality education, training and experience. Although training and education are often used interchangeably, there are important distinctions. You educate for understanding and train for performance. A person attends university in order to obtain an in-depth understanding of the subject matter they are studying, and later are trained in specific techniques to accomplish the task at hand. One without the other is a recipe for disaster. An educated individual with no training has limited useful skills for solving problems. A well-trained, but uneducated person could be a useful technician who can follow instructions, but would most likely be unable to interpret images to resolve problems. Proper application of the scientific method requires the scientist to be both well educated and well trained.

The seven case examples contained in this chapter exemplify the fact that images are real data, too, which are valuable for solving problems.

Reference

Merriam-Webster. (n.d.). *Subjective*. In Merriam-Webster.com dictionary. Retrieved September 5, 2022, from https://www.merriam-webster.com/dictionary/subjective

3.1 Mesothelioma Linked to Asbestos

John A. Reffner, Ph.D.[1] and Brooke W. Kammrath, Ph.D.[2,3]

[1] John Jay College of Criminal Justice, New York, NY, USA
[2] University of New Haven, West Haven, CT, USA
[3] Henry C. Lee Institute of Forensic Science, West Haven, CT, USA

Fibrous asbestos minerals have been used for over 4500 years, and for nearly half that time it has been suspected that there is a link between its exposure and respiratory illnesses (Lee & Selikoff, 1979; McCulloch & Tweedale, 2007; King, 2022; Meredith, n.d.; Mesothelioma Hope Team, 2022). People exposed to asbestos mining, fabrication of materials, or its use would develop impairment of their lungs. This was considered to be similar to black lung disease and silicosis, where foreign bodies (e.g., coal and silica) are trapped in the lungs and restrict breathing. Although asbestos-related illnesses were known since the times of Pliny the Younger (61–112 A.D.), it was not recognized as a serious concern by the medical community until the rapid growth of the asbestos industry and its widespread use in the 19th century. The long induction period between initial exposure and onset of impaired breathing (between ten and sixty years), combined with the fact that not all who are exposed to asbestos develop cancer, contributed to this delay in appreciating the magnitude of the asbestos problem.

Asbestos is the term for six fibrous silicate minerals (chrysotile, amosite, crocidolite, tremolite, anthophyllite, and actinolite). Chrysotile is classified as a serpentine mineral, while the other five are in the amphibole mineral class based on their morphologies. Asbestos minerals are excellent thermal and electrical insulators which are heat resistant, and thus have found numerous uses as construction (e.g., plaster, drywall, insulation) and fireproofing (e.g., fire blankets, firefighter clothing) materials. Other uses of asbestos minerals included cigarette filters and chemical laboratory equipment (e.g., heating screens and Gooch crucibles). By the early 20th century, materials containing asbestos minerals were ubiquitous.

The asbestos problem was first perceived as an industrial issue. In England, the United States, and Canada, a number of people raised concerns over this problem. The focus in the early 20th century was on asbestosis, a chronic lung disease marked by lung tissue scarring and shortness of breath . In 1918, the U.S. Bureau of Labor Statistics released a report stating there was an unusually high early mortality rate for those who worked in the asbestos industry (Meredith, n.d.). In 1929, Dr. W.E. Cooke published "Asbestos dust and the curious bodies found in pulmonary asbestosis" in the *British Medical Journal*, where he reported on a case study of the death of an asbestos worker in 1924 where he was asked "Did the deceased's occupation cause or contribute to her death?" This seminal paper followed his 1927 *Journal of the Royal Microscopical Society* publication "Pneumokoniosis due to asbestos dust" (Cooke and Hill, 1927". Cooke used both brightfield and stereomicroscopy, as well as X-ray diffraction, to examine what he called "curious bodies" which "consist of central nuclei of asbestos spicules upon which colloidal aggregates of blood proteins, plus, possibly soluble fractions of asbestos, and in the case of chrysotile workers an iron salt, have been absorbed and molded by currents in the bronchi and alveoli" (Cooke, 1929, p. 579). His ultimate conclusion was that "the curious bodies, if found in any numbers, are pathognomonic of pulmonary asbestosis." In 1930, Dr. E.R.A. Merewether, a British barrister and physician, essentially extended the work of Cooke by performing a clinical examination of hundreds of asbestos workers and found that one in four suffered from asbestosis. This initiated regulations of the asbestos industry in the United Kingdom, but not the United States or Canada.

Shortly thereafter, in 1934, the link between cancer and prolonged exposure to asbestos was first shown in research (Meredith, n.d.). When extensive epidemiological studies became available, the correlation of both asbestos and lung cancer were clearly demonstrated. For over 30 years, hundreds of research studies were performed and published that established a correlation between prolonged exposure to asbestos and lung cancer. This included warnings to the general population from a physician at the National Cancer Institute, Dr. W.C. Heuper, who indicated that prolonged asbestos exposure via mining, manufacturing, or daily environmental exposure through finished building materials caused both asbestosis and cancer. Despite this, the asbestos industry failed to warn workers of this significant health risk and no regulations were implemented (Meredith, .n.d.).

The issue remained. Correlation does not prove causation, and there had yet been shown a direct cause-and-effect relationship between asbestos minerals and lung cancer. In the mid-1950s, Dr. Irving J. Selikoff was a General Practice physician in New Jersey, and the Asbestos Workers Union hired him to treat their members. After noticing an unusually high occurrence of pleural (lung) mesothelioma in this population, he focused his career on investigating the link between asbestos and lung cancer. In 1964, Dr. Selikoff and his research team published the seminal work "Asbestos exposure and neoplasia" in the *Journal of the American Medical Association*, which not only confirmed the known correlation of cancer with asbestos exposure, but extended the danger to anyone working in the asbestos mining, manufacturing, and construction industries (e.g., builders, electricians, plumbers, carpenters), their family members, and those who live in the vicinity of asbestos mines and factories. Thus, the authors effectively branded asbestos as a significant environmental health risk which required immediate attention, and followed it with subsequent publications (Selikoff et al., 1965 and 1968). In 1966, Dr. Selikoff founded and became the director of the Environmental and Occupational Health Division of the Mount Sinai School of Medicine of the City University of New York. While at Mount Sinai, he and other scientists validated prior research by applying light and electron microscopical analysis and microchemistry to the investigation of lung tissue and isolated "asbestos bodies" which consisted of fibrous materials encased in an iron-rich protein material (Langer et al., 1972; Figure 3.3). The iron-rich protein coating had previously been a major barrier to identifying the composition of the fibrous core using instrumental methods. For example, cupper X-rays excite iron atoms generating characteristic iron X-rays which mask diffraction patterns. The significance of the work by Selikoff and his team was that they conclusively identified the fibrous cores as being asbestos minerals, thus essentially finding the "epidemiological correlation of asbestos fiber exposure and morbidity-mortality data" (p. 724).

While the evidence was clear that asbestos was a health problem for decades, it was not regulated by the US government until the 1970s. This is in contrast to the UK where regulations were imposed upon the asbestos industry in 1931 and workers

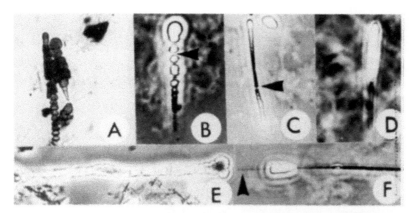

Figure 3.3 Characteristic asbestos bodies as viewed with normal light (A) and phase contrast light (B–F). All bodies were exposed by ashing histological sections. Segmentation of coating is common (A, B, C, and E) with bulbous ends singly terminated (C and E). Occasionally, no or little segmentation is visible (D) or is nonuniform in size and distribution (B and F). Arrowheads in B, C, and F indicate areas along the bodies where the core is exposed. Bodies range in size from 10 μ (D) to 25 μ (E). (Langer et al., 1972/The Histochemial Society, p. 724).

diagnosed with asbestosis were compensated like they would be for any job-related injury. The first US government regulation of asbestos was in a 1970 amendment to the Clean Air Act which extended its control of air pollution to include industrial sources including asbestos. In 1972, the Occupational Safety and Health Administration (OSHA) became involved with the asbestos problem, and they developed regulations for asbestos workers. As a result of these regulations, there was a need to identify microscopic asbestos pollution in the workplace and other public locations. Microscopy solved this problem.

Walter C. McCrone at McCrone Associates, Inc. in Chicago, Illinois, developed a method using a polarized light microscope (PLM) and employing dispersion staining technology to specifically identify fibrous asbestos and their chemical composition as one of the six asbestos minerals. Dispersion staining provides a specific pseudo-coloring of the different fibrous asbestos minerals when they are mounted in fluid media of known refractive index (as seen in Figure 3.4). This method was adopted by OSHA as a standard method for identification of bulk asbestos materials in the workplace and traces in the environment, and was quickly used throughout the world. In addition, other microscopic methods were employed, including phase contrast and transmission electron microscopy, for asbestos identification and quantitation in air samples.

Once the government began looking for asbestos in the environment, it was found everywhere. Since asbestos was used in numerous types of building materials during the construction boom mid-century, asbestos was found in many public buildings including schools. In 1986, the Asbestos Hazard Emergency Response Act (AHERA) was implemented under the Toxic Substance Control Act. This required all private and public schools to be inspected for the presence of asbestos. Consequently, asbestos identification became a big business because of the great concern over its public health effects.

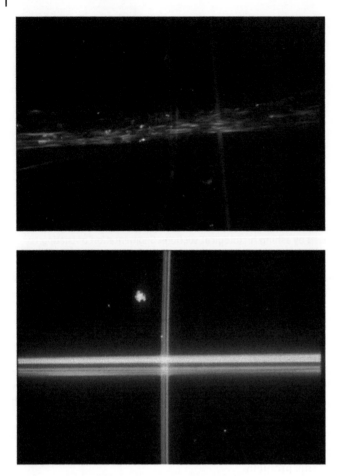

Figure 3.4 Central stop dispersion staining of asbestos fibers: chrysotile in 1.550 high dispersion (HD) oil (top) and Amosite in 1.670 HD oil (bottom).

In the 1980s, the EPA attempted to phase out asbestos-containing materials; however, the US courts overturned this and thus asbestos is still in use although highly regulated. In 2002, the last US asbestos mine was closed which effectively ended a once booming industry. As of 2020, 55 countries have asbestos bans from all products. Notably absent from this list of countries are the US, Canada, China, and Russia. Asbestos is still mined in Canada, Russia, and Africa, and imported into the US. Worldwide, billions of dollars continue to be spent yearly on health care for asbestos-related diseases and asbestos remediation.

Microscopy was vital not only to understanding the health problems caused by asbestos, it also was and still is the key tool for its identification and subsequent remediation. Every asbestos laboratory continues to use microscopy to identify fibrous asbestos in the environment.

References

Cooke, W. E. (1929). Asbestos dust and the curious bodies found in pulmonary asbestosis. *British Medical Journal*, *2*(3586), 578. doi: 10.1136/bmj.2.3586.578

Cooke, W. E., & Hill, C. F. (1927). Pneumokoniosis due to asbestos dust. *Journal of the Royal Microscopical Society*, *47*(3), 232–238. https://doi.org/10.1111/j.1365-2818.1927.tb01413.x

King, D. (2022, May 24). *The history of asbestos - importing, exporting & worldwide use.* Mesothelioma Center - Vital Services for Cancer Patients & Families. Retrieved September 5, 2022, from https://www.asbestos.com/asbestos/history

Langer, A. M., Rubin, I. B., & Selikoff, I. J. (1972). Chemical characterization of asbestos body cores by electron microprobe analysis. *Journal of Histochemistry & Cytochemistry*, *20*(9), 723–734.

Lee, D. H., & Selikoff, I. J. (1979). Historical background to the asbestos problem. *Environmental Research*, *18*(2), 300–314. https://doi.org/10.1016/0013-9351(79)90107-5

McCulloch, J., & Tweedale, G. (2007). Science is not sufficient: Irving J. Selikoff and the asbestos tragedy. *New Solutions*, *17*(4), 293–310.

Meredith, N. (n.d.). *The history of asbestos: Timelines of when asbestos was first used.* Mesothelioma Help Cancer Organization. Retrieved September 5, 2022, from https://www.mesotheliomahelp.org/asbestos/history

Mesothelioma Hope Team. (2022, January 4). *Banning asbestos in the United States - history and facts.* MesotheliomaHope.com. Retrieved September 5, 2022, from https://www.mesotheliomahope.com/legal/legislation/asbestos-bans

Selikoff, I. J., Churg, J., & Hammond, E. C. (1964). Asbestos exposure and neoplasia. *JAMA*, *188*, 22–26.

Selikoff, I. J., Churg, J., & Hammond, E. C. (1965). Relation between exposure to asbestos and mesothelioma. *New England Journal of Medicine*, *272*(11), 560–565. doi: 10.1056/NEJM196503182721104

Selikoff, I. J., Hammond, E. C., & Churg, J. (1968). Asbestos exposure, smoking, and neoplasia. *JAMA*, *204*(2), 106–112. doi: 10.1001/jama.1968.03140150010003

3.2 Talc Case

John A. Reffner, Ph.D.[1] and Brooke W. Kammrath, Ph.D.[2,3]

[1] John Jay College of Criminal Justice, New York, NY, USA
[2] University of New Haven, West Haven, CT, USA
[3] Henry C. Lee Institute of Forensic Science, West Haven, CT, USA

A link between cancer and talc, commonly known as baby powder, has been suspected since the 1970s (Cramer et al., 1982). It is believed that the use of talc in feminine hygiene products increases the risk of ovarian, cervical, and other types of cancer. There are several common methods for genital exposure to talc, including use as a deodorant, using condoms with talc as a lubricant, and the powdering of diaphragms and sanitary napkins. Similar to the asbestos problem, the failure to determine a definitive link between talc and cancer is exasperated by the long induction period between exposure and diagnosis combined with the fact that not all women who used talc products for feminine hygiene develop cancer.

The history of the talc problem is similar to that of the asbestos problem previously detailed; however, there are four important distinctions that have prevented widespread acceptance of this association. First, there are no reports of other talc exposures (i.e., mining, manufacturing, or environmental) causing cancer. Second, although the chemistry of talc and asbestos are similar, their morphologies are different with talc being plate-like while asbestos minerals are fibrous. Third, various studies (International Agency for Research on Cancer, 2016) have shown that there is variability in the increased risk of developing ovarian cancer, between 30 and 60 percent, creating questions about the reliability of the data. Last, early studies showed the presence of asbestos in talc, thus contributing to a thought that it was the asbestos causing the cancers as opposed to the talc itself.

Since the 1970s, producers of talc products have separated out the asbestos, thus creating asbestos-free talc. Unlike with asbestos, there is no evidence of talc producers having an early awareness of this problem nor attempting to deceive the general public. Still, epidemiological studies of women who use asbestos-free talc for feminine hygiene continue to develop cancer.

Early in the 1970s, microscopy was used to identify talc particles in cervical and ovarian cancer (Henderson et al., 1971). Scanning electron microscopy equipped with energy dispersive X-ray spectroscopy (SEM-EDX) was used in this first paper from 1971 to identify talc in tissue sections. Today, PLM is used by pathologists to detect the highly birefringent talc crystals in cancerous tissue sections (Figure 3.5) and elemental analysis is used for confirmation. It must be recognized that there are other birefringent materials present in tissue, such as collagen fibers and calcifications, thus the presence of anisotropic materials alone does not immediately indicate the presence of talc. The existence of talc in cancerous tissues, as detected by microscopy, is highly suggestive of a causal link between talc and cancer. The issue is now being adjudicated in the courts.

In 2009, the first talcum powder civil lawsuit was filed against Johnson & Johnson by a woman who developed ovarian cancer in 2006. Although she was offered a deal for $1.3 million, she declined. In 2013, she won her liability case; however, no compensation was awarded. Numerous lawsuits followed, but it was not until a 2016 civil case in Missouri that a verdict included monetary damages ($10 million in actual damages and $62 million in punitive damages). This case was appealed (Mueller, 2016; Stempel, 2016). In the summer of

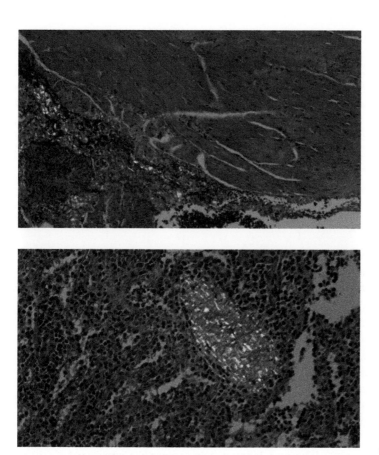

Figure 3.5 PLM photomicrographs of tissue sections showing the presence of anisotropic particles consistent with talc. The bottom image shows numerous red blood cells which each measure ~7 μm in diameter.

2018, Gail L. Ingham and 21 other women (or their families) filed a petition in St. Louis City Circuit Court against Johnson & Johnson, alleging claims for strict liability, negligence and other torts, specifically alleging they developed ovarian cancer after using their talc products. In July 2021, after more than 6 weeks of trial and testimony from over 30 witnesses, the jury returned a verdict finding the Defendants liable on all claims. The jury awarded $550 million in compensatory damages ($25 million to each plaintiff) and $4.14 billion in punitive damages. In June 2020, a Missouri appeals court upheld this verdict, but reduced the damages to $2.1 billion (the compensatory damages were unchanged but the punitive damages were cut to $1.6 billion) (Ingham v. J&J, ED-107476; Feeley, 2020). Johnson & Johnson further appealed this verdict to the US Supreme Court; however, it ended on June 1, 2021, when the highest court rejected the request to review this case (Trager, 2021). As of the writing of this book, tens of thousands of lawsuits against Johnson & Johnson persist in the court systems. Although the microscopy is definitive in determining the presence of talc minerals in cancer tumors, there is no direct evidence that is the cause of the tumor formation. It is still uncertain whether producers of talc products are legally responsible.

In May of 2020, Johnson & Johnson reported that they would stop the sale of talc-based products in the United States and Canada. This was based on declining sales of these products which they characterized as due to misinformation rather than proof of danger. They cited a January 2020 *Journal of the American Medical Association* article (O'Brien et al., 2020) which found that there was no "statistically significant association between use of powder in the genital area and ovarian cancer" based on a study of over 250,000 women. This demonstrates the obvious conflict between science and the courts. In the court of public opinion, talc-based products were found guilty of causing cancer. As a result, these products are no longer available in the United States and Canada.

References

Cramer, D. W., Welch, W. R., Scully, R. E., & Wojciechowski, C. A. (1982, Jul 15). Ovarian cancer and talc: A case-control study. *Cancer, 50*(2), 372–376. https://doi.org/10.1002/1097-0142(19820715)50:2<372::AID-CNCR2820500235>3.0.CO;2-S

Feeley, J. (2020, June 23). *J&J talc verdict cut in half to $2.1 billion by state court.* Bloomberg.com. Retrieved September 5, 2022, from https://www.bloomberg.com/news/articles/2020-06-23/court-cuts-4-7-billion-j-j-talc-verdict-to-2-1-billion

Henderson, W. J., Joslin, C. A. F., Griffiths, K., & Turnbull, A. C. (1971, March). Talc and carcinoma of the ovary and cervix. *Journal of Obstetrics and Gynecology of the British Commonwealth, 78*, 266–272. https://doi.org/10.1111/j.1471-0528.1971.tb00267.x

Ingham v. J&J, ED-107476, Missouri court of appeals for the eastern district (St. Louis).

International Agency for Research on Cancer. (2016). Carbon black, titanium dioxide, and talc. *IARC monographs on the evaluation of carcinogenic risk to humans, 93*, pp. 193e275.

Mueller, A. (2016, November 7). Johnson & Johnson seeks to move talc lawsuits out of St. Louis. *St. Louis Business Journal.* Retrieved September 5, 2022, from https://www.bizjournals.com/stlouis/news/2016/11/07/johnson-johnson-seeks-to-move-talc-lawsuits-out-of.html

O'Brien, K. M., Tworoger, S. S., Harris, H. R., Anderson, G. L., Weinberg, C. R., Trabert, B., ... Wentzensen, N. (2020). Association of powder use in the genital area with risk of ovarian cancer. *Jama, 323*(1), 49–59. doi:10.1001/jama.2019.20079

Stempel, J. (2016, February 24). *J&J must pay $72 million for cancer death linked to Talcum Powder: Lawyers.* Reuters. Retrieved September 5, 2022, from https://www.reuters.com/article/us-johnson-johnson-verdict-idUSKCN0VW20A

Trager, R. (2021, June 4). *US Supreme Court will not hear J&J Talc Appeal.* Chemistry World. Retrieved September 5, 2022, from https://www.chemistryworld.com/news/us-supreme-court-will-not-hear-jandj-talc-appeal/4013805.article

3.3 Ford Pinto Case

John A. Reffner, Ph.D.[1] *and Brooke W. Kammrath, Ph.D.*[2,3]

[1] *John Jay College of Criminal Justice, New York, NY, USA*
[2] *University of New Haven, West Haven, CT, USA*
[3] *Henry C. Lee Institute of Forensic Science, West Haven, CT, USA*

In 1972, a Ford Pinto hatchback stalled on a freeway and, upon being rear-ended, burst into flames. This resulted in the death of the female driver, Lily Gray, and severe burns to her 13-year old passenger, Richard Grimshaw. As a result, a personal injury tort case was filed against the car manufacturer in 1978 in Orange County, California, and the decision was affirmed by a California appellate court in 1981 (*Grimshaw v. Ford Motor Company*, 119 Cal. App.3d 757, 174 Cal.Rptr. 348). The court of public opinion was dramatically influenced by an article in Mother Jones magazine entitled "Pinto Madness" which accused the Ford Motor Company of putting a low dollar value on human life when performing a cost-benefit analysis on whether to make safety improvements in their vehicles (Dowie, 1977; Schwartz, 1990).

There were two components of the accident that were focused on at trial: (1) the reason for the initial stalling of the car, and (2) the cause of the car fire after the rear-end collision. For the latter, there were design flaws that included a flimsy bumper and poor placement of the fuel tank in the rear of the vehicle, among others. It was argued by the plaintiff that had the Ford company addressed the vehicle design with a focus on safety, at a cost of approximately $9 per vehicle, the car would not have caught fire. The defense argued that there were little regulations on cars being rear ended in the early 1970s, the Ford Pinto met crash testing requirements at the time which occurred at ~20 mph, and that this accident and resulting fire occurred due to the excessive speed (~50 mph) at which the Pinto was struck. It must be noted that the cause of the fire cannot be answered by microscopy, and thus will not be focused on here. However, microscopy was essential for addressing the initial question about the cause of the stalling of the Pinto.

The question at trial was the status of the carburetor float. A carburetor is used to control the fuel-air ratio in an engine, and was common in pre-1990 vehicles, when it was replaced by electronic fuel injection systems. The carburetor float is in a side reservoir that acts as a miniature fuel tank, that regulates the flow of fuel to the air stream. When the fuel level in this reservoir is low, the carburetor float triggers an increase in fuel flow fuel to the engine. Conversely, when the fuel level is high, the flow is restricted. In this case, the plaintiff argued that "the carburetor float had become so saturated with gasoline that it suddenly sank, opening the float chamber and causing the engine to flood and stall" (Grimshaw v. Ford Motor Company, 1981). However, the car was traveling at a high rate of speed, thus the fuel was flowing, which questions the plausibility of this argument. Still, this "heavy" carburetor float became a key item of evidence which required examination.

The carburetor float used in the Ford Pinto consisted of two black plastic foam structures and a metallic hinge, as shown in Figure 3.6. It must be noted that when the carburetor float was removed from the carburetor of the Ford Pinto, the two black plastic foam pieces each broke in half at the hinge insertion location. This resulted in the carburetor float evidence consisting of four black plastic foam pieces and the metallic hinge.

Figure 3.6 Ford Pinto Carburetor Float.

The plaintiff hired an engineer to examine the carburetor float. At trial, he testified that he determined that the carburetor float was "heavy" due saturation with gasoline. On cross examination, he described his "analysis" as consisting of him going to his "goody box" and finding four metal nuts that seemingly weighed the same as the metal hinge from the carburetor float. He never actually weighed the metal nuts, only compared them in his hands by placing the four nuts in his left hand and the hinge in his right hand. He then attached one nut to each of the four pieces of the plastic carburetor float, and suspended them in gasoline. The pieces sank to the bottom. Based on this, he concluded that the floats were defective. However, when questioned on cross examination, he responded that he never claimed any scientific certainty for his observations or testing.

The defense expert was a material scientist with a background in both polymers and microscopy. He was provided with the four pieces of the carburetor float. He noted that the black plastic foam pieces had both smooth outer surfaces and rough fractured surfaces. The rough fractured surfaces provided the examination of the internal structure of the foam polymer. There are two types of foam: open-cell and closed-cell. In creating a foam structure, gas bubbles are formed surrounded by the polymer. In a molded structure, such as the carburetor float, the gas bubbles can be forced together forming a network of angular structures (rhombic dodecahedrons with 12-faced surfaces) creating a closed-cell foam. However, if the bubbles burst as the material is cured, the spherical bubbles retain their shape but create open channels, hence an open-cell foam. Microscopically, these appear as spheres. On the outer walls of the molding, the polymer forms a continuous solid barrier. To determine if a foam is open-cell or closed-cell, microscopic examination of the internal structure is required. In this case, reflected light darkfield microscopy was used, as shown in Figure 3.7. The angularity present in the image in Figure 3.7 A and B, making rhombic dodecahedrons, is evidence of closed-cell foam. The carburetor float's structure was observed to have an angular morphology characteristic of a closed-cell foam. The polymer

Figure 3.7 The typical internal structure of a molded closed-cell foam (A and B) and an open-cell foam (C), as viewed by reflected light darkfield microscopy.

(A)

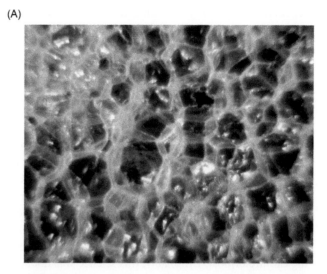

(B)

(C)

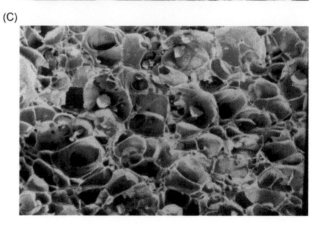

base of this float is an ebonized nitrile rubber, which is extremely impenetrable by any fluid, including gasoline. Gasoline could not diffuse through the polymer wall nor would it have been able to saturate into the foam because there are no open channels for it to flow into. The combination of the ebonized nitrile rubber surface combined with the closed-cell foam internal structure is evidence that this was not a defective carburetor float.

Despite the conclusive evidence that the carburetor float was not the cause of the accident, the plaintiffs won the case with the jury awarding them $127.8 million in damages ($125 million in punitive damages and $2.8 million in compensatory damages), which was the largest ever in US product liability and personal injury cases until the mid-1990s. Subsequently, the judge reduced the punitive damages to $3.5 million. This decision was affirmed upon appeal. This case also was the impetus for additional safety regulations for the automotive industry and stricter guidelines for crash testing.

The science described in this case demonstrates some facts that were presented at trial but did not make it into the public domain. There is no readily available written record describing the testimony of the defense witness who performed microscopical analysis of the carburetor float. Ultimately, this science was not appreciated by the jury and ignored in the court ruling and subsequent magazine and law review articles analyzing this case. Was justice served?

References

Dowie, M. (1977). Pinto madness. *Mother Jones, September/October*, 18–32.

Grimshaw v. Ford Motor Company. (1981). (119 Cal.App.3d 757, 174 Cal.Rptr. 348)

Schwartz, G. T. (1990). The Myth of the Ford Pinto Case. *Rutgers Law Review, 43*, 1013.

3.4 Uncovering a Moose Hair Cover-up

John A. Reffner, Ph.D.[1] and Brooke W. Kammrath, Ph.D.[2,3]

[1] John Jay College of Criminal Justice, New York, NY, USA
[2] University of New Haven, West Haven, CT, USA
[3] Henry C. Lee Institute of Forensic Science, West Haven, CT, USA

One night, a hit-and-run accident occurred on a New York State Parkway resulting in the death of the bicyclist who was struck. The bicyclist was hit by a car and thrown up onto the hood and windshield of the car. Upon investigating local car repair shops, a detective found a vehicle with damage on the passenger side fender, hood and windshield, consistent with striking the bicyclist. When questioned about the cause of the damage, the vehicle owner stated that he had struck a deer. On examination of the vehicle, there was evidence of blood and animal hair which appeared to substantiate the driver's claim. To validate the driver's alibi, the detective collected the animal hairs and sent them to the forensic laboratory for microscopical identification.

The forensic laboratory reported that the unknown animal hairs were moose hairs and not deer hair as claimed by the driver. There are several microscopic features of animal hairs that enable them to be classified and identified. These include the medullary index, the medullary pattern, surface (or cuticle) scale pattern, and root morphology (Deedrick & Koch, 2004; Petraco & Kubic, 2003). Moose hairs can be identified by their wide lattice medullary pattern and mosaic scale pattern in the shape of large irregular polygons (Figure 3.8a) in the basal region of the hair shaft. In comparison, although deer hair also contains a wide lattice medullary pattern with a mosaic cuticle pattern, their scales are more round and resemble the scales of a fish (Figure 3.8b).

(a)

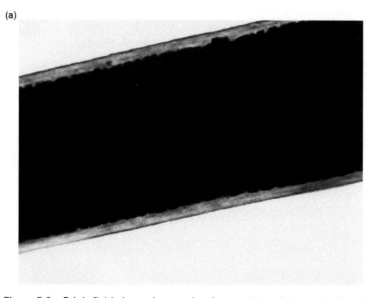

Figure 3.8 Brightfield photomicrographs of moose (a) and deer hair (b) as viewed with a 20-times objective.

(b)

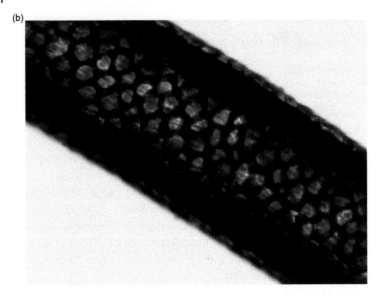

Figure 3.8 (Cont'd)

After receiving the forensic hair examination report, the detective knew that something was amiss because moose are not known to live in southern New York. The detective then went to the home of the driver for additional questioning. When the detective entered the suspect's home, he saw a large moose head mounted on the wall. The detective then confronted the driver with the forensic examiner's conclusion of the recovered hair being from a moose, thus refuting the driver's alibi. The driver immediately broke down and confessed to the hit-and-run crime, and his attempted cover-up which included removing hairs from his taxonomized moose and putting them on the damaged vehicle. The driver thought he was being clever by creating a plausible alibi with the moose hair; however, he did not understand the power of microscopy to perform animal hair identifications.

References

Deedrick, D., & Koch, S. (2004). Microscopy of hair part II- A practical guide and manual for animal hairs. *Forensic Science Communications, 6*(3).

Petraco, N., & Kubic, T. (2003). *Color atlas and manual of microscopy for criminalists, chemists, and conservators*. CRC Press.

3.5 Carbon Black and Tire Rubber Problems

John A. Reffner, Ph.D.[1] and Brooke W. Kammrath, Ph.D.[2,3]

[1] *John Jay College of Criminal Justice, New York, NY, USA*
[2] *University of New Haven, West Haven, CT, USA*
[3] *Henry C. Lee Institute of Forensic Science, West Haven, CT, USA*

Carbon black has been used as a colorant since prehistoric times. It is the black residue produced by the incomplete combustion reaction of hydrocarbons, making it a readily available pigment for ink and paints centuries before its properties were understood. With the advent of automobiles, carbon black became a ubiquitous additive used in the manufacturing of tires (Whitby et al., 1954; Garvey and Freese, 1942). Early tires used iron oxide pigments to both color and reinforce the rubber; however, by the 1920s, carbon black was the principle pigment for automotive tires. Today, carbon black is a multi-billion-dollar industry with it being an important pigment used in numerous products, including rubber, plastics, inks, toners, paints, and coatings (Industry Experts, 2019). Seventy percent of the carbon black produced is used in the tire industry. Carbon black not only is a relatively inexpensive extender material that colors the tires black, it also increases the rubber's tensile strength, weatherability, wear resistance, and conduction of heat away from the tread and belt areas thus extending the life of the tires. It has been estimated by Jack Koenig in his book *Spectroscopy of Polymers* that an automobile tire produced without carbon black would last less than 5000 miles (Koenig, 1992), which is approximately ten times less than tires with carbon black. An often underappreciated feature of carbon black is the role that the size and morphology of the individual particles have on the properties of the final product. Although other technologies are used to study and identify the chemistry of the carbon black pigments, such as gas chromatography-mass spectrometry and Raman microspectroscopy, electron and light microscopy remain a valuable tool.

Light microscopy is an efficient method for providing information on the particle size and morphology of carbon blacks which are critical for determining the final properties of the material. The microscope has been used for solving problems in the rubber industry since the early 1900s. Dr. Raymond Peck (R.P.) Allen was a student of Emile Chamot at Cornell University, and subsequently became employed at the B.F. Goodrich Tire and Rubber Company ("Goodrich"). He was in charge of the Works Technical microscopy laboratory in Akron, Ohio. Allen was a very successful, respected, valued employee and proud microscopist. In his 1943 article entitled "Technical Microscopy in the Rubber Industry," Allen eloquently states:

> For supplying information not obtainable by the usual methods, the microscope is useful and often essential; not only will it disclose details which would otherwise be unseen, but by its unique methods of examination measurements of certain properties of materials can be readily secured which are not otherwise obtainable.

(p. 219)

In this work, Allen described several microscopical approaches to determine particle size, shape, and dispersion of pigment particles in rubber raw materials and finished products. There are many different types of carbon blacks which vary in particle size and morphologies which cause differences in their observed color and chemical properties. Identifying the specific type of carbon black in a material is critical. Allen used darkfield microscopy at various magnifications, which included stereomicroscopy, to investigate samples containing carbon black. The simplest method for carbon black identification is to take small fragments of a known standard and a questioned carbon black in rubber, place them in contact with each other, side-by-side, in a drop of mineral oil and between two microscope slides with some pressure (Figure 3.9). Upon examination of the interface between the standard and the unknown, they can be compared to determine the identity of the unknown. If no distinct interface is visible, a match is made. Other than identification, the microscope is valuable for quality control of the raw materials and factory processing, to check pigment characteristics and ensure complete dispersion of the particles. Further, microscopy was shown to solve problems that occur at the rubber manufacturing facility. The microscopical analytical techniques described by Allen to solve problems and characterize materials were extended to other industries that use carbon black. Two case examples demonstrate the value of microscopy in the rubber industry.

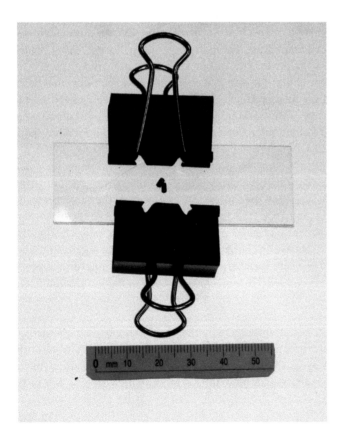

Figure 3.9 Two carbon black-filled rubber samples mounted side-by-side in an immersion oil, between two glass slides. By examining the interface, the color and specific shade of black are able to be compared to determine if the two samples are indistinguishable.

3.5.1 Goodrich's Airplane De-icers

B.F. Goodrich was the inventor of pneumatic airplane wing de-icers in the 1920s. Along with Goodrich tires, these de-icers were used in 1927 by Charles Lindbergh in his historic transatlantic flight in the Spirit of St. Louis from New York to Paris. Subsequently, Goodrich became the leading producer of de-icing technology. In 1957, their supremacy in the de-icer market was threatened by a competitor who introduced a more practical electrically heated de-icer. This consisted of a thin piece of rubber with two electrical wires on either side. Goodrich's research group had attempted to make an electrically heated de-icer using conductive carbon black, but had failed to do so. They found that the process of stretching the rubber around the leading edge of the wing caused the electrical resistance to become too high, thus no longer conducting electricity and failing to function as an airplane de-icer. The competitor had solved this problem, to the chagrin of the Goodrich executives. The Goodrich microscopy department was tasked with identifying the carbon black used in this new electrically heated de-icer. The microscopist accepted this assignment thinking that it would be no problem, as he had been routinely identifying carbon blacks following Dr. Allen's procedures. However, after comparing the sample to all carbon blacks in the Goodrich collection, no identification was made. Fortuitously, the microscopist was a member of the Rubber division of the American Chemical Society (ACS), and subscribed to their *Rubber Chemistry and Technology* journal. When he returned home from work that evening, the latest issue was in his mailbox, which featured an article by Polley and Boonstra (1957) entitled "Carbon Blacks for Highly Conductive Rubber." This article described the chemical properties of various new carbon blacks and contained a graph showing their change in electrical resistance versus elongation. The Vulcan SC carbon black showed a low resistance over a large elongation (i.e., stretching), which was exactly the property of the unknown sample that enabled it to work as a de-icer. The next day, the microscopist went to the Goodrich factory supply room to inquire if they had any of the Vulcan SC carbon black. The clerk said that a small sample had been received for testing, but had not yet been used. A reference sample, which consisted of the Vulcan SC carbon black in rubber, was prepared for the microscopist to investigate. When this sample was prepared side-by-side with the unknown carbon black specimen from the competitor, and examined using darkfield microscopy, it became immediately clear that they were the same. The darkfield image of the two samples with no visible interface was the data proving the common identity of the carbon black. The microscopist reported this to the Goodrich management, who were excited and relieved. As a result, Goodrich was able to develop their own electrically heated de-icer, using the Vulcan SC carbon black, which kept them at the forefront of de-icing technologies.

3.5.2 A Problem with Retreaded Airplane Tires

Retreading airplane tires is a common practice. When an airplane lands, the tread rubber is abraded and requires replacement. However, in one case, there was a failure of the newly retreaded tires that occurred when the airplane was taking off. Although it did not negatively affect the flight, it was alarming to the pilot and ground crew. The unexpected malfunction, which consisted of unusual lack of adhesion of the tread rubber to the tire carcass, was of great concern. As a result, the retreaded tires were brought to the Goodrich microscopy lab for examination. The first question was whether the adhesive had been properly applied to the interface between the carcass and the retread. The carcass of a tire is the portion that maintains its pneumatic properties (e.g., air pressure), and consists of fibers and a different polymer (often a butyl rubber). The microscopist made a cross section of the tire, and examined it microscopically using an ultraviolet light source. This was used to observe the adhesive, which fluoresced when illuminated with ultraviolet light. The presence of the adhesive was obvious, thus it was present and not the source of the tire failure. Next, the tread rubber was examined using darkfield reflected light microscopy. The microscopist observed sand grains in the rubber, which should not have been present because the highest quality natural rubber was required for retreading airplane tires. The sand causes an unusual internal stress in the rubber that prevents adhesion with the tire carcass, and was thus determined to be the cause of the problem. The next dilemma was to identify the source of the sand grains in the rubber. Although there are always some particulate matters from the environment in natural rubber, it is graded based on the quantity and size of these impurities. High-quality natural rubber contains minimal amounts of particulates. The microscopist inquired about the origin of the natural rubber, and obtained a sample of the raw material that was used to make the tire treads. Upon examining the raw natural rubber used in the airplane retreads, the excessive sand grains were obvious when observed with a stereomicroscope. This was reported to management, and resulted in the development of a more rigorous process for grading raw materials used in the retreading of airplane tires.

References

Allen, R. P. (1942). Technical microscopy in the rubber industry.

Garvey, B. S., & Freese, J. A. (1942). Effect of carbon blacks in synthetic tire compounds. *Industrial & Engineering Chemistry, 34*(11), 1277–1283. https://doi.org/10.1021/ie50395a006

Industry Experts. (2019). *Carbon black – A global market overview.* Industry Experts Report CP026. Retrieved September 5, 2022, from https://industry-experts.com/verticals/chemicals-and-materials/carbon-black-a-global-market-overview

Koenig, J. L. (1992). *Spectroscopy of polymers (ACS professional reference book).* Wiley-VCH.

Polley, M. H., & Boonstra, B. B. S. T. (1957, *March*). Carbon blacks for highly conductive rubber. *Rubber Chemistry and Technology, 30*(1), 170–179. https://doi.org/10.5254/1.3542660.

Whitby, G. S., Davis, C. C., & Dunbrook, R. F. (Eds.). (1954). *Synthetic rubber.* J. Wiley.

3.6 Optical Microscopy Takes Center Stage: Melamine in Pet Food

Mark R. Witkowski, Ph.D. and John B. Crowe

U.S. Food and Drug Administration, Forensic Chemistry Center, Trace Examination Section, Ohio, USA

3.6.1   Introduction

Whether walking or driving, when we come to a street corner or intersection we usually stop and look both ways before crossing or pulling out into the road. If not and we just walk across the street or pull out into an intersection, we may get lucky and not get hit or we may not be so lucky. This analogy can be applied to problem solving. Just taking a sample, preparing it, and analyzing is like walking into an intersection without stopping and looking. We may or may not get lucky and find the answer to our problem, and in some instances, we may even create more problems. However, if we stop and look at our sample prior to preparation and analysis, we may find information which will lead to a more straightforward quicker answer to the problem. As the complexity of the problem increases, a multi-disciplinary approach (i.e., multiple instrumental techniques and methods) may be necessary to solve the problem with optical microscopy being part of this approach (Bloise & Miriello, 2018; Lanzarotta et al., 2012).

In the last 20 years there has been a revolution in analytical instrumentation allowing for lower detection limits and the ability to detect a wider range of analytes. Nowhere is this more apparent than in the area of mass spectrometry where highly sensitive exact-mass-determining instruments coupled with chromatography or direct-analysis attachments can detect very low levels of different analytes. However, being able to detect an analyte(s) at a very low level may not be enough to solve complex problems. These highly sensitive instrumental techniques require some form of sample preparation (e.g., dissolved or extracted) prior to analysis. Sample matrix information and the original form of the analyte (e.g., morphology, crystallinity) are altered or lost with these sample preparation methods. This lost information may be critical to solving a problem.

Since its invention, the optical microscope has continued to be an important analytical tool used in scientific research and problem solving (Clay & Court, 1932; Hajdu, 2002; Wollman et al., 2015). Being able to magnify and view a sample with no preparation allows one to derive a large amount of information about a sample very quickly. The very fact that a sample is examined in its solid state, means the microscope can provide the scientist with important information about a sample which is lost in sample preparation. Visual information such as the number of particle types, homogeneity, morphology, crystallinity, sizes, and color can be very useful to the investigating scientist (McCrone et al., 1997). The optical microscope can provide information on what caused the problem but also can provide information on the how and why the problem occurred.

Economically motivated adulteration (EMA) is the intentional adulteration of a food product or products for financial gain (Everstine et al., 2013; Moyer et al., 2017). EMA may include adulterating a high value food with low cost ingredients or passing off a low-quality food as high quality by adulterating it with low cost ingredients. In either case, EMA can cause a serious economic impact to the legitimate manufacturer but more importantly, it poses a very serious public health risk since the adulterant used may have unintended consequences to the unsuspecting consumer. In 2007, a voluntary recall of/for pet food began when a series of adverse events occurred after cats and dogs consumed certain brands and manufacturing lots of wet pet food (Dobson et al., 2008; Osborne et al., 2008; Reimschuessel et al., 2008; Rovner, 2008). As the number of animal illnesses and deaths increased, the recall increased in size resulting in one of the largest pet food recalls in U.S. history. The affected animals exhibited the same adverse symptoms after consuming wet pet food. Although the suspect pet food consumed represented different brands, all were manufactured at a single facility. Samples of the suspect pet food were submitted to the US FDA Forensic Chemistry Center (FCC) and other laboratories for analysis to determine what was causing the deaths (Dobson et al., 2008; Litzau et al., 2020; Osborne et al., 2008; Reimschuessel et al., 2008; Rovner, 2008).

Using mass spectrometry, melamine and s-triazine compounds (ammeline, ammelide, and cyanuric acid) were detected and identified in the suspect wet pet food samples (Figure 3.10A) (Dobson et al., 2008; Litzau et al., 2020; Rovner, 2008). Melamine and the other s-triazine compounds are used to manufacture a variety of products, and based on past toxicology information, these individual compounds posed a low health risk (Dobson et al., 2008; Osborne et al., 2008). Although the "what" question had been answered, the detection of melamine and s-triazine compounds still did not answer three other very important questions. "How did melamine and s-triazine compounds find their way into the suspect wet pet food samples?" "Why were these compounds there at all?" and "Why were these compounds causing so much harm to animals?" Optical microscopy and other solid-state analysis techniques such as Fourier transform infrared (FT-IR) spectroscopy and Raman spectroscopy played a major role in answering those three remaining questions (Dobson et al., 2008; Osborne et al., 2008; Reimschuessel et al., 2008).

3.6.1.1 How Did Melamine and the Other S-triazine Compounds Find Their Way into the Suspect Wet Pet Food Samples?

Based on the initial instrumental analysis findings and discussions with the manufacturer, the lots and brands of wet pet food associated with the animal illnesses and deaths coincided with a change in supplier of two ingredients (Dobson et al., 2008; Osborne et al., 2008;

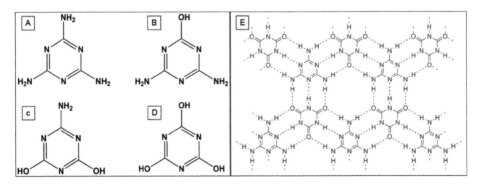

Figure 3.10 The molecular structures of (A) melamine, (B) ammeline, (C) ammelide, (D) cyanuric acid, (E) hydrogen-bonded complex of melamine cyanurate.

Rovner, 2008). The two ingredients, wheat gluten and rice protein concentrate, were now being supplied by a new company from China. In the manufacture of pet food, wheat gluten is used as a binder and rice protein concentrates are used as a vegetable protein source (Day, 2011; Hoogenkamp et al., 2016). Control samples of wheat gluten and rice protein concentrate from the previous supplier and suspect samples of these two ingredients from the new supplier were provided to the FCC and others for analysis (Dobson et al., 2008; Litzau et al., 2020; Osborne et al., 2008; Rovner, 2008).

The control and suspect wheat gluten and rice protein samples were initially screened using mass spectrometry. As with the wet pet food samples, mass spectrometry detected melamine and s-triazine compounds (ammeline, ammelide and cyanuric acid) but at higher amounts than in the wet pet food samples. Based on these results, it was decided to examine the raw ingredient samples using optical microscopy (stereomicroscopy and polarized light microscopy (PLM)). Based on what was observed using optical microscopy, further analysis would be done using FT-IR spectroscopy. These solid-state analysis techniques were used to determine in what form the melamine and s-triazine compounds were present in the raw ingredients and if the mass spectrometry results had missed something in the screening work.

The first step was to visually compare the suspect and control samples using stereomicroscopy and document any and all differences observed between the samples (Figure 3.11). It was found the suspect samples were not visually consistent with the control samples. The control wheat gluten sample consisted of a homogenous tan colored powder (Figure 3.11A) while the suspect wheat gluten sample consisted of a heterogenous tan-to-white powder with multiple particle types (Figure 3.11B). Three distinct particles types were observed in the suspect wheat gluten. Type 1 particles consisted of off-white to gray irregular-shaped particles, type 2 particles consisted of clear to opaque irregular-shaped particles and type 3 particles consisted of white irregular-shaped particles (Figure 3.11B). The control rice protein concentrate sample consisted of a homogenous light tan colored powder (Figure 3.11C) and the suspect rice protein concentrate sample consisted of a heterogenous white powder with multiple particle types (Figure 3.11D). Three distinct particles types were also observed in the suspect rice protein concentrate sample. Type 1 particles consisted of tan irregular shaped particles and type 2 particles consisted of clear to opaque irregular-shaped particles and type 3 particles consisted of white irregular-shaped particles (Figure 3.11D). Visually the particle types 2 and 3 from the suspect wheat gluten were consistent with particles types 2 and 3 found in the rice protein concentrate.

The different particles types were physically isolated for additional analysis using FT-IR spectroscopy. The type 1 and 2 particles isolated from the wheat gluten were consistent with melamine. The type 3 particles were consistent with a melamine-like compound but did not match the FT-IR spectra of any of the standard spectra of melamine, ammeline, ammelide, or cyanuric acid. Mass spectrometry confirmed the melamine in type 1 and 2 particles but identified a mixture of melamine, ammeline, ammelide, and cyanuric acid in the type 3 particles. The same FT-IR and mass spectrometry results were obtained for types 2 and 3 particles isolated from the rice protein concentrate.

By visually examining the samples, valuable information was obtained that was important to the investigation. The suspect wheat gluten and rice protein concentrates samples were visually different from the control samples. Because of the visually differences

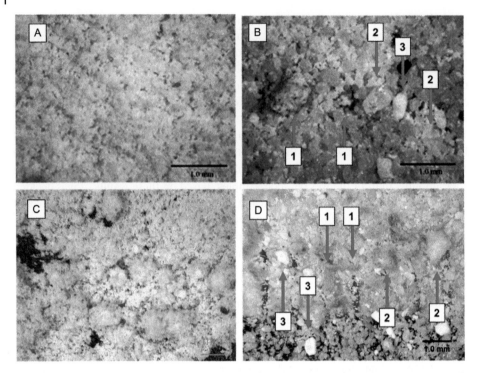

Figure 3.11 Reflected light stereomicroscope photomicrographs of (A) control wheat gluten, (B) suspect wheat gluten containing three different unknown particles types, (C) control rice protein concentrate, (D) suspect rice protein concentrate containing three different unknown particle types.

and different particles types, stereomicroscopy was able to rapidly screen additional raw material ingredients. The melamine and s-triazine compounds were present as individual particles which appeared to have been added to the matrix and a unique particle type which appeared to be a melamine-like compound.

3.6.1.2 Why Were These Compounds There at All?

Melamine and the s-triazine compounds found in the wet pet food and the raw material ingredients contain many nitrogen atoms (Figure 3.10). A review of the literature found that these compounds and specifically melamine had been studied for potential use as a protein supplement due to the amount of nitrogen present in their structures (Dobson et al., 2008; Osborne et al., 2008; Reimschuessel et al., 2008; Rovner, 2008). The total protein content of wheat gluten and rice protein concentrates are typically determined as total nitrogen content using the Kjeldahl method (Regenstein & Regenstein, 1984). The total nitrogen content is then used to calculate the amount of protein present in the raw materials. The Kjeldahl method is nonspecific and cannot differentiate nitrogen from protein or another source. Because of the high number of nitrogen atoms present in melamine and the other s-triazine compounds, they could be easily added to the wheat gluten and rice protein concentrates to increase the total nitrogen content. Based on this information, further questions were raised–whether the wheat gluten was wheat gluten or if the rice protein concentrates were rice protein concentrate?

The easiest way to answer these questions was to perform further optical microscopical examinations on the suspect wheat gluten and the rice protein concentrate samples. Powder from both the control and suspect samples were mounted on glass slides with cover slips using a 50:50 mixture of glycerin: water (n ~ 1.40) (Hoyt, 1934). Both suspect and control samples were visually examined with a PLM and compared to each other using both plane polarize light and crossed polarized light. Figure 3.12 shows the comparison of the control samples to the suspect samples viewed in plane polarized light. The suspect wheat gluten and rice protein concentrate samples are again visually different from the control samples. In fact, it was determined that the suspect wheat gluten and rice protein concentrate samples were composed mainly of wheat flour with very little wheat gluten or rice protein concentrate present. The suspect samples (Figure 3.12B and 3.12E) contained particle types (plant material, wheat starch) consistent with the particle types found in a control wheat flour (Figure 3.12C and 3.12F) (Greenish, 1923; Winton & Barber Winton, 1945; Winton and Moeller, 1906).

Based on the microscopic analyses, it appeared that the wheat gluten and rice protein concentrate ingredients from the new supplier were nothing more than wheat flour with melamine and s-triazine compounds added to boost the nitrogen content so they would appear to contain high amounts of protein. These pseudo wheat gluten and rice protein concentrate ingredients were then used to manufacture the wet pet food brands and lots associated with the animal illnesses and deaths.

At this point we were able to determine the "what," "how," and "why" but we still were unable to answer two final questions. Eventually, microscopy would lead the way to the identification of the melamine-like compound and provide us the answer.

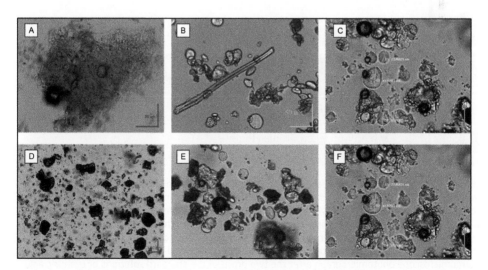

Figure 3.12 Plane-polarized light images of (A) control wheat gluten, (B) suspect wheat gluten, (C) control wheat flour, (D) control rice protein, (E) suspect rice protein, (F) control wheat flour.

3.6.1.3 What Was the Melamine-like Compound and What Role if Any Did It Play in Causing Harm to the Animals?

The individual melamine and the s-triazine compounds have been shown to have limited toxicity. However, the observation and detection of a melamine-like compound (type 3 particles) raised the possibility that this compound may have higher toxicity and possibly be responsible for causing the harm to animals. The mass spectrometry results of the type 3 particles isolated from the suspect raw materials showed they contained melamine and s-triazine compounds, but the mass spectral data did not show the presence of this additional melamine-like compound observed by FT-IR. Based on the mass spectral data, we knew all four compounds were present in the melamine-like compound (type 3) particles. We decided to try and generate the type 3 particle in the laboratory using a standard of melamine and mixing it with the other s-triazine compounds. The laboratory scientists produced mixtures which were then analyzed by optical microscopy, FT-IR, and Raman spectroscopy.

Standards of powder forms of melamine, ammeline, ammelide, and cyanuric acid were visually examined using PLM (using both plane and crossed polarized light) and analyzed by FT-IR and Raman microscopy. Saturated solutions of each standard were prepared in reverse-osmosis water. The standard solutions were mixed together in various combinations using the two-drop method. When performing this two-drop microcrystalline test, mixture reactions are performed on a microscope slide and directly observed using the microscope so that conclusions can be drawn about the observed reactions. A drop of one standard solution was placed

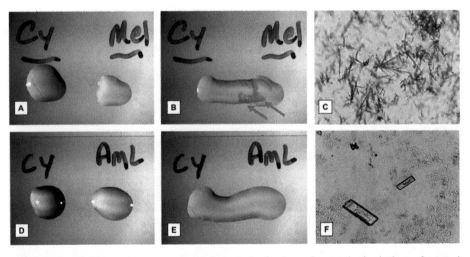

Figure 3.13 (A) Stereomicroscope photomicrograph of a drop of a standard solutions of cyanuric acid (Cy) and melamine (Mel), (B) stereomicroscope photomicrograph of instantaneous crystal formation (red arrows) after the two drops of the standard solutions cyanuric acid and melamine are drawn together, (C) plane polarized light photomicrograph of the resultant melamine: cyanuric acid crystals formed after drying, viewed with a 20-times objective, (D) stereomicroscope photomicrograph of a drop of a standard solutions of cyanuric acid (Cy) and ammeline (Aml), (E) stereomicroscope photomicrograph after the two drops of the standard solutions of cyanuric acid and ammeline are drawn together, no instantaneous crystal formation, (F) plane polarized light photomicrograph of the resultant ammeline: cyanuric acid crystals formed after drying, viewed with a 20-times objective.

on a glass slide then a second drop of a different standard solution was placed on the slide opposite the first drop (Figure 3.13). The two drops were visually examined as they were pulled together and allowed to dry. Spontaneous crystals formed when the melamine and cyanuric acid solutions were drawn together (Figure 3.13B). However, no spontaneous crystal formation was observed for the ammeline and cyanuric acid mixture (Figure 3.13E), or the ammelide and cyanuric acid mixture. A final experiment involved preparing a mixture of melamine, ammeline, and ammelide and then mixing them with cyanuric acid. Spontaneous crystallization was observed when the mixture solution and cyanuric acid solutions were drawn together.

For each standard mixture, the dried crystals were examined using PLM (with both plane polarized and crossed polarized light) and then analyzed using FT-IR and Raman microscopy. All four mixtures produced unique crystal shapes. The melamine: cyanuric acid mixture formed birefringent acicular single crystals which agglomerated into groups (Figure 3.14A1, A2). The ammeline: cyanuric acid mixture formed low-to-medium birefringent trapezoidal crystals (Figure 3.14B1, B2). The ammelide: cyanuric acid mixture formed medium birefringent single euhedral, rectangular crystals (Figure 3.14C1, C2). It should be noted that to obtain crystallization of ammeline: cyanuric acid and ammelide: cyanuric acid crystals the solutions were dried in a vacuum oven. The forth mixture of a melamine: ammeline: ammelide: cyanuric acid consisted of conglomerate of crystals dominated by the acicular crystals of melamine: cyanuric acid. These crystals appear to look like sea urchins, a solid center with protruding needles. In all cases, the resultant crystals from the four standard mixtures could be differentiated by PLM, FT-IR, and Raman microscopy.

The FT-IR spectra of the melamine: cyanuric acid crystals and the melamine: ammeline: ammelide: cyanuric acid crystals were consistent with the FT-IR spectra collected of the melamine-like compound (type 3) particles and tentatively identified as melamine cyanurate. This was confirmed with a melamine cyanurate standard. When melamine and cyanuric acid are mixed together in solution, they form the hydrogen-bonded complex melamine cyanurate (Figure 3.1E) (Hajdu, 2002; Wollman et al., 2015).

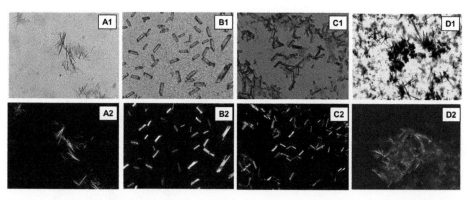

Figure 3.14 Photomicrographs of (A1) melamine: cyanuric acid crystals, viewed in plane (A2) and crossed polarized light, (B1) ammeline: cyanuric acid crystals viewed in plane (B2) and crossed polarized light, (C1) ammelide: cyanuric acid viewed in plane (C2) and crossed polarized light, (D1) melamine: ammelide: ammelide: cyanuric acid viewed in plane-(D2) and crossed polarized light. All viewed with a 20-times objective.

Based on the toxicology information, the ingestion of melamine, the s-triazine compounds, or melamine cyanurate should not have been an issue. The hydrogen-bonded complex of melamine cyanurate should have disassociated upon ingestion leaving only melamine and cyanuric acid. However, the spontaneous crystal formations observed when mixing standard mixtures of melamine and cyanuric acid and the mixture of the three s-triazines with cyanuric acid was very intriguing. Could the spontaneous crystal formation have played a role in the harm caused to the animals? Bladder stones were found in the animals that had ingested the affected food and the FT-IR spectra of these stones were consistent with melamine cyanurate. This indicated that the melamine cyanurate was passing through the digestive system in to the urinary system. In addition to the bladder stones, cross sections of kidney tissue from animals which had died revealed crystal formations in the individual renal tubes (Figure 3.15A).

Cross sections of kidney tissue containing these crystal formations were received from the FDA's Center for Veterinarian Medicine (CVM). These cross sections were examined using optical microscopy and the crystal formations were visually consistent with the crystal formations observed when melamine, ammeline, and ammelide were mixed with cyanuric acid (Figure 3.6B). These crystals were then analyzed using Raman microscopy and confirmed the presence of melamine cyanurate.

The analysis of the crystal formations in the kidney tissue samples answered last the question, "what role if any did it play in causing harm to the animals?" In the end, the presence of the melamine, cyanuric acid, and melamine cyanurate proved to be extremely harmful when ingested by the animals. The hydrogen-bonded complex melamine cyanurate would indeed disassociate upon ingestion. However, once the melamine and cyanuric acid passed through the digestive system and entered the urinary system the hydrogen-bonded complex would reform. Over time the melamine cyanurate crystals would form crystals and block the renal tubes of the kidneys resulting in kidney failure, then death (Dobson et al., 2008; Osborne et al., 2008; Reimschuessel et al., 2008).

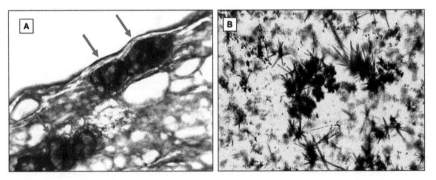

Figure 3.15 (A) Plane polarized light photomicrograph of a portion of cat kidney tissue cross section viewed with 40-times objective. The melamine: cyanuric acid crystals observed in the tubes (red arrows), (B) plane polarized light photomicrograph of melamine: ammelide: ammelide: cyanuric acid crystals viewed with 20-times objective.

3.6.2 Conclusion

Although the individual compounds melamine, ammeline, ammelide, and cyanuric acid were found in the original suspect wet pet food samples and raw materials using mass spectrometry, optical microscopy was needed to show that the raw ingredients were fraudulent and that they contained individual particles of melamine the hydrogen-bonded complex, melamine cyanurate. Optical microscopy provided the visual evidence which demonstrated the formation of spontaneous crystals upon mixing melamine with cyanuric acid. It also was able to confirm that the optical and morphological properties of the crystals prepared in the laboratory were consistent with the crystal formations observed in the kidney renal tubes.

This was a perfect example of how using the fundamentals of optical microscopy combined with other instrumental techniques helped solve a complex problem. In an age where sophisticated laboratory instrumentation takes center stage, the use of a microscope has been relegated to the back stage. However, there are times (like this one) where the modern microscope designed long ago by Robert Hooke can keep pace and even perform better than the most sophisticated laboratory instrument. So next time you have a complex problem to solve, stop and look. You might be surprised that the answer is right before your eyes.

Acknowledgments

The work presented in this paper represents only a small portion of work performed by the U.S. FDA. We would like to recognize Jonathan Litzau, Kevin Mulligan, and Rick Flurer who performed the mass spectrometry analysis and all the employees at FCC, CVM, and other FDA groups who were involved in analyzing these and other samples. Also, we would like to recognize all the private companies, universities, and state laboratories who were working either independently or collaborating with us. It was this collective effort which made possible the quick resolution of this crisis.

References

Bloise, A., & Miriello, D. (2018). *Geosciences, 8*, 1. https://doi.org/10.3390/geosciences8040133

Clay, R. S., & Court, T. H. (1932). *The history of the microscope*. Charles Griffin and Company.

Day, L. (2011). Wheat gluten: Production, properties and application, handbook of food proteins. In G. O. Phillips & P. A. Williams (Eds.), *Woodhead publishing series in food science, technology and nutrition* (pp. 267–288). Woodhead Publishing.

Dobson, R. L., Motlagh, S., Quijano, M., Cambron, R. T., Baker, T. R., Pullen, A. M., Regg, B. T., Bigalow-Kern, A. S., Vennard, T., Fix, A., Reimschuessel, R. (2008). *Toxicological Sciences, 106*(1), 251. https://doi.org/10.1093/toxsci/kfn160

Everstine, K., Spink, J., & Kennedy, S. (2013). *Journal of Food Protection, 76*(4), 723.

Greenish, H. G. (1923). *The microscopical examination of foods and drugs*. P. Blakiston's Son & Co.

Hajdu, S. I. (2002). *Annals of Clinical & Laboratory Science, 32*(3), 309.

Hoogenkamp, H., Kumagai, H., & Wanasundara, J. P. D. (2016). Rice protein and rice protein products. In S. Nadathur, J. P. D. Wanasundara, & L. Scanlin (Eds.), *Sustainable protein sources* (pp. 47–65). Academic Press.

Hoyt, L. F. (1934). *Industrial & Engineering Chemistry*, *26*(3), 329. https://doi.org/10.1021/ie50291a023

Lanzarotta, A., Crowe, J. B., Witkowski, M., & Gamble, B. M. (2012). *Journal of Pharmaceutical and Biomedical Analysis*, *67-68*, 22. https://doi.org/10.1016/j.jpba.2012.04.023

Litzau, J. J., Mercer, G. E., & Mulligan, K. J., U.S. FDA laboratory information bulletin (LIB) 4423: Melamine and related compounds, https://www.fda.gov/food/laboratory-methods-food/laboratory-information-bulletin-lib-4423-melamine-and-related-compounds (accessed 27 April 2020)

McCrone, W. C., McCrone, L. B., & Delly, J. G. (1997). Polarized light microscopy, McCrone Research Institute.

Moyer, D. C., DeVries, J. W., & Spink, J. (2017). *Food Control*, *71*, 358. https://doi.org/10.1016/j.foodcont.2016.07.015

Osborne, C. A., Lulich, J. P., Ulrich, L. K., Koehler, L. A., Albasan, H., Sauer, L., & Schubert, G. (2008). *Veterinary Clinics: Small Animal*, *39*, 1. https://doi.org/10.1016/j.cvsm.2008.09.010

Regenstein, J. M., & Regenstein, C. E. (1984). *Protein quantitation, food protein chemistry, an introduction for food scientists* (pp. 90–108). Academic Press.

Reimschuessel, R., Gieseker, C. M., Miller, R. A., Ward, J., Boehmer, J., Rummel, N., Heller, D. N., Nochetto, C., de Alwis, G. H., Bataller, N., Andersen, W. C. (2008). *American Journal of Veterinary Research*, *69*(9), 1217. https://doi.org/10.2460/ajvr.69.9.1217

Rovner, S. L. (2008). *Chemical and Engineering News*, *86*(19), 41.

Winton, A. L., & Barber Winton, K. (1945). *The analysis of foods*. John Wiley & Sons, Inc.

Winton, A. L., & Moeller, J. (1906). *The microscopy of vegetable foods*. John Wiley & Sons.

Wollman, A. J. M., Nudd, R., Hedlund, E. G., & Leake, M. C. (2015). *Open Biology*, *5*, 1. https://doi.org/10.1098/rsob.150019

3.7 Characterization of Foreign Particulate in Pharmaceuticals

Glass Corrosion

Richard S. Brown, M.S.

MVA Scientific Consultants, Inc., Duluth GA, USA

One of the more difficult types of analysis involves the appearance of small, unwanted particles (less than one millimeter) in manufactured products. Isolation and characterization of these particles is best achieved using microscopy. Using microscopical techniques, the particle shape, fracture surfaces, brittleness, plasticity, color, optical properties, and inclusions can be characterized without destroying the sample. In our laboratory we routinely use a combination of polarized light microscopy (PLM), reflected darkfield microscopy, reflected differential interference contrast (DIC) microscopy, Fourier transform infrared microscopy (FT-IR), scanning electron microscopy coupled with energy dispersive X-ray spectrometry (SEM-EDX), and confocal Raman microscopy (CRM) to characterize small particles isolated from manufactured products. One type of particle is particularly interesting because it results from the corrosion of a material that is generally not associated with corrosion: glass. (Hubbrad and Hamilton, 1941; Schaut et al., 2020)

Glass is used as a container for foods, medicines, beverages, and chemicals because it tends to be resistant to reacting with the product (Schaut and Weeks, 2017). Occasionally, glass finds itself in contact with liquid environments that can, over time, cause it to corrode. The corrosion process involves ion exchange between the solution and the glass, the dissolution of the glass network, pitting, stable film formation, and surface layer exfoliation (Hench, 1985). In extreme cases of glass container corrosion, the number of exfoliated particles, or lamella, can be substantial causing a clear solution to become cloudy. The presence of these glass corrosion products is unacceptable for anything intended for human consumption or injection. Thus their detection and identification is critical for ensuring product quality. (Zuucato, n.d.)

Microscopy is uniquely suited for the characterization of this exfoliated glass corrosion product. The characterization process is broken down into two phases. The first phase involves isolating the particles from the solution. Sample preparation is performed by documenting the sample condition and photographing any particle present prior to filtering the container contents. For small clear glass containers, a camera system attached to a stereobinocular microscope is adequate to capture images of particulate in the container prior to isolation from the container. Water-based products can be vacuum filtered through a polycarbonate membrane filter apparatus set up within an ISO 5 clean bench. The filter is examined using a combination of stereobinocular microscopy using reflected and oblique light, and reflected DIC microscopy. Photomicrographs are taken to document the types of particles present on the filter. Particle types are determined by morphology, color, and size.

Subsamples of the filter can be cut and mounted on a suitable substrate for analysis of the particulate by FT-IR, CRM, and SEM-EDX. The glass corrosion particles are the breakdown product of the glass they originated from. The resulting corrosion particles or lamella are thin film residues that cling to the filter. Physical isolation by picking or lifting representative particles from the membrane is extremely difficult. Once the filter membrane has dried, the thin glass corrosion residues tend to bond to the filter preventing their removal. They do not retain the same chemical composition of the glass from where they originated, and they are nearly invisible on the membrane filter. Occasionally, a stack of several particles may show faint interference colors giving a slight indication that they are present. Using reflected DIC microscopy, glass corrosion particles are visible as thin film fragments. SEM-EDX analysis of the filter requires careful examination of the filter surface using low accelerating voltages to locate and analyze these particles. Many of the particles are so thin that they are electron transparent resulting in the electron beam passing through the particle.

Figure 3.16 shows glass corrosion product on a polycarbonate membrane filter that has been isolated by filtration from a glass container. Subsampling a section of the filter allows the particles to be imaged and analyzed directly on the filter surface by SEM-EDX using a low accelerating voltage (2–10 kV). The corrosion product shown in Figure 3.16 is from a glass container that showed severe corrosion when examined by SEM (Figure 3.17). Containers where corrosion is present as determined by SEM examination may result in minimal glass corrosion product isolated on the filter membrane that can be overlooked using reflected brightfield microscopy. A reflected light DIC microscope can be used to provide superior imaging (Figure 3.18) of thin, sparse corrosion product particles that can be easily missed using reflected brightfield imaging alone (Figure 3.19). To confirm that the particles are the result of glass corrosion the internal diameter of the vial must be examined for evidence of corrosion.

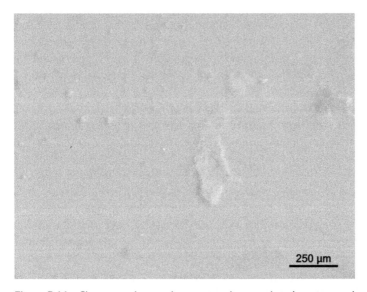

250 µm

Figure 3.16 Glass corrosion product captured on a polycarbonate membrane filter using vacuum filtration. Image was taken with a stereobinocular microscope with reflected brightfield illumination.

Figure 3.17 Secondary electron image of the internal diameter of a glass container exhibiting severe corrosion. Exfoliation is occurring with the gel layer fracture visible at the surface.

Figure 3.18 Reflected DIC image of glass corrosion thin film fragments on a polycarbonate membrane surface.

The second phase involves examination of the internal diameter of the container. The container is thoroughly rinsed with water for injection (WFI) then examined using stereobinocular microscopy with reflected and oblique light, and reflected DIC microscopy. Reflected DIC can be used at low magnification (50× to 100×) to inspect the internal diameter of the container. The DIC reflectance objectives are not corrected for use with a coverslip nor are

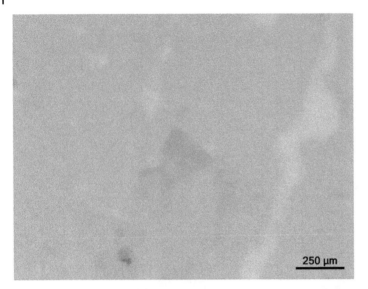

250 µm

Figure 3.19 Reflected brightfield image of glass corrosion thin film fragments on a polycarbonate membrane surface.

they corrected for the glass wall thickness of the container. The thickness of the glass causes light to scatter and results in a poor image. Fortunately, most glass containers can be easily sectioned using a diamond saw. We usually cut the containers in half, then half again to get a section that can be examined with reflected DIC and using the SEM.

The corrosion process in glass containers favors degradation of the glass surface in stressed areas produced during manufacture. Such areas include the heel near the base of a vial and the shoulder of the container. The liquid surface or the fill line of the product should be marked and examined for corrosion. The fill line is the area where a corrosive material can concentrate and/or react with any oxygen present by being at the air interface within the container. Many times, gross observation of the container will reveal a sheen or haze in corroded areas in the container. Haze or interference colors in a glass container may be documented by careful oblique illumination and a good quality stereobinocular microscope (Figure 3.20). Once the internal diameter of the cut section is examined by reflected DIC (Figure 3.21) areas of suspected corrosion are marked for additional examination using SEM. A thin conductive gold coating applied to the internal diameter of the glass section after DIC examination aids in imaging the nonconductive glass surface. Once an area of possible corrosion is located, imaging at different magnifications will reveal the presence of corrosion (Figures 3.22 A–C and 3.23). Using an accelerating voltage of 5 kV, remaining organic residue, salts, or product that may be adhering to the internal diameter of the container can be characterized by the amount of carbon or the presence of elements not in the glass. A reference EDS spectrum can be obtained from a fresh fracture surface near the cut edge of the glass section. Comparison of the elemental composition of the glass container to residues on the internal diameter of the glass helps in determining whether the residue is product or corrosion. Product

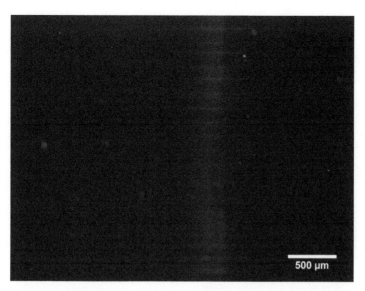

Figure 3.20 Reflected light image using oblique illumination of haze in a glass container caused by glass corrosion.

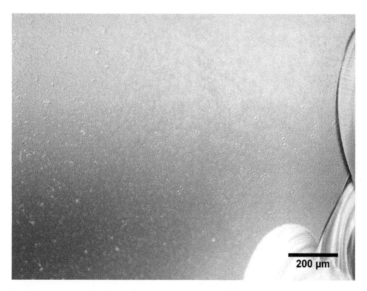

Figure 3.21 DIC image of haze on internal diameter of glass container caused by glass corrosion.

residues can be removed by rinsing with deionized water and methanol. Stubborn residues can be removed by ultrasonication with a mild detergent followed by rinsing with deionized water and methanol.

To determine a glass container's potential corrosion resistance new containers from different manufacturers may be exposed to solutions and elevated temperature that will produce corrosion under the controlled conditions. In this way, a container's resistance to

(A)

Figure 3.22 Secondary electron images of haze on internal diameter of glass container caused by glass corrosion at 250× (A), 1500× (B), and 12,000× (C) magnifications.

(B)

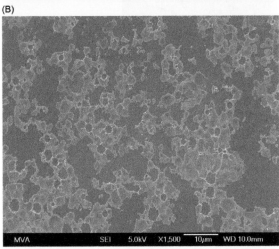

(C)

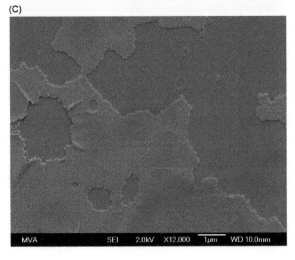

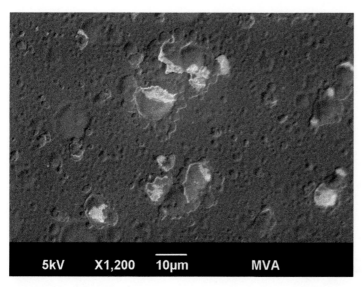

5kV   X1,200   10µm   MVA

Figure 3.23   Secondary electron image of glass corrosion pitting with exfoliation of glass corrosion surface.

these specific solutions can be evaluated (Panighello & Pinato, 2020). Removal of the liquid and inspection of the internal diameter of the container by DIC and SEM is necessary to confirm the presence of and the extent of any corrosion to the glass. For characterizing the degree of corrosion, new unfilled reference containers should be inspected by DIC and SEM to determine the condition of the glass surface with special attention to any potential corrosion sites (high stress areas) introduced during manufacturing. For cases of severe corrosion, reference containers are not necessary. Through thorough examination and comparison of the test and reference containers' internal surfaces a trained microscopist can classify the corrosion as not present, early onset, present or severe. Unfortunately, the use of corrosive test solutions is not a reliable predictor of a glass—product interaction. Periodic evaluation of the glass container with the product present is the only way to determine product compatibility with the glass during storage.

To microscopically examine a corrosion reaction surface, a thin cross section from a single corrosion pit was made using a focused ion beam microscope. This microscope is a dual beam scanning microscope equipped with an electron beam for imaging and EDS analysis, and an ion beam for imaging, sample trenching, and sample polishing. The thin section process can be observed during the trenching and polishing steps while they are being performed. Figures 3.24A and 3.24B show SEM images of a corrosion pit selected for thin section. The trenching process cuts away the glass surrounding the area selected for lift out and polishing using a gallium ion beam (Figure 3.25). The resulting thin section shows the profile of the corrosion pit in cross section. The reaction zone is visible in this scanning transmission electron microscope (STEM) image (Figure 3.26). The transmission electron microscope (TEM) image shows the reaction zone in detail (Figure 3.27).

(A)

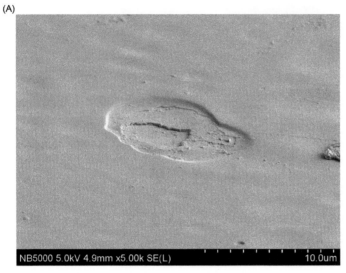

NB5000 5.0kV 4.9mm x5.00k SE(L)   10.0um

(B)

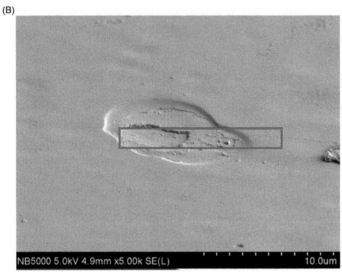

NB5000 5.0kV 4.9mm x5.00k SE(L)   10.0um

Figure 3.24 Secondary electron image of glass corrosion pit showing exfoliation (A) and the area selected for thin sectioning outlined with a red box (B).

A combination of reflected brightfield microscopy, reflected DIC microscopy and SEM can determine the extent of glass corrosion in glass containers used for pharmaceuticals. Microscopy is uniquely suited for this type of analysis because the exfoliated corrosion particles and the glass surface they originated from can be investigated directly resulting in photomicrographs that document the condition of the glass surface, the corrosion if present, and in many cases, the severity of the corrosion.

NB5000 5.0kV 4.9mm x2.50k SE(L)   20.0um

Figure 3.25 Secondary electron image of focused ion beam milling to define section for lift out and polishing.

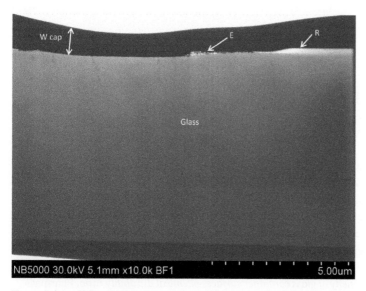

W cap

Glass

E

R

NB5000 30.0kV 5.1mm x10.0k BF1   5.00um

Figure 3.26 STEM image of thin section after lift out and focused ion beam polishing. Dark layer "W" is tungsten cap sputtered onto to sample to protect surface during ion beam milling. Reaction zone "R" leading to exfoliation "E" is indicated.

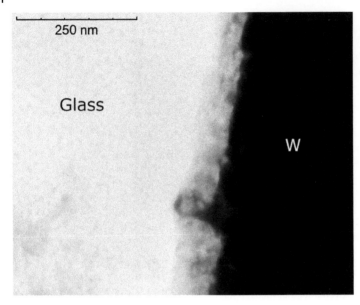

250 nm

Glass

W

Figure 3.27 TEM image of area "E" in Figure 3.26. Image has been rotated clockwise 90 degrees.

Acknowledgment

Thank you to the Clemson University Electron Microscopy Facility for preparing the focused ion beam lift out.

References

Hench, L. L. (1985). Glass Corrosion. In A. F. Wright & J. Dupuy (Eds.), *Proceedings of the NATO advanced study institute on glass...current issues* (pp. 551–554). Martinus Nijhoff. NATO AST Series E: Applied Sciences, No. 92.

Hubbrad, D., & Hamilton, E. H. (1941, August). *Studies of the chemical durability of glass by an interferometer method, research paper RP1409.* Part of journal of research of the national bureau of standards (Vol. *72*). U.S. Department of Commerce. https://nvlpubs.nist.gov/nistpubs/jres/27/jresv27n2p143_A1b.pdf

Panighello, S., & Pinato, O. (2020, March-April). Investigating the effects of the chemical composition on glass corrosion: A case study for type I vials. *PDA Journal of Pharmaceutical Science and Technology, 74*(2), 185–200. https://doi.org/10.5731/pdajpst.2019.010066

Schaut, R., Murphy, K., Kramer, D., Chapman, C., & Flynn, C. (2020, 14 July). *A new method to quantify the heterogeneity of glass container surfaces-chemical durability ratio,* Pharma@corning.com. https://www.corning.com/media/worldwide/cpt/documents/CPT_Valor_WhitePaper_CDR_V5_08-2020_FINAL.pdf

Schaut, R. A., & Weeks, W. P. (2017, July). Historical review of glasses used for parenteral packaging. *PDA Journal of Pharmaceutical Science and Technology, 71*(4), 279–296. https://journal.pda.org/content/71/4/279

Zuucato, D. (n.d.) *Glass delamination: Risks, reality and regulatories.* PDA Presentations. Accessed February 1, 2021, from https://www.pda.org/docs/default-source/website-document-library/chapters/presentations/metro/glass-delamination-risks-reality-and-regulatories.pdf?sfvrsn=2dc5a38e_6

4

The Microscope as a Compass

Look Before You Act

Introduction

Microscopy is capable of being a navigational aid, or compass, to plot the course for successful problem solving. The first step of any scientific inquiry should be to perform a comprehensive visual examination of a specimen. Observation is the first step of the scientific method for several important reasons. Making visual observations is an efficient process that can be documented and is nondestructive. Further, observations of a sample, using both macroscopic (visible to the naked eye) and microscopic (visible with a microscope) methods, provide information which are helpful in solving problems. This was famously demonstrated by Louis Pasteur in his examination of wine sediments, specifically paratartaric acid, using a polarized light microscope. Images can help not only to determine the next steps of analysis, but also prevent one from performing unnecessary testing and wasting resources. In many cases, if observations are not made at the outset of a scientific investigation, the opportunity to do so is lost forever.

The six cases presented in this chapter demonstrate the ability of microscopical examinations to direct a variety of diverse scientific investigations with the goal of solving problems.

Solving Problems with Microscopy: Real-life Examples in Forensic, Life and Chemical Sciences, First Edition.
Edited by John A. Reffner and Brooke W. Kammrath.
© 2024 John Wiley & Sons Ltd. Published 2024 by John Wiley & Sons Ltd.

4.1 Hair Extension Case

John A. Reffner, Ph.D.[1] and Brooke W. Kammrath, Ph.D.[2,3]

[1] John Jay College of Criminal Justice, New York, NY, USA
[2] University of New Haven, West Haven, CT, USA
[3] Henry C. Lee Institute of Forensic Science, West Haven, CT, USA

In 2011, at the behest of an unscrupulous attorney, a woman was asked to purchase hair extensions that were labelled "100% human hair." The attorney then sent samples of these blond (Figure 4.1) and black hair extensions to a private laboratory for DNA analysis to determine if they were human or animal hairs, or neither. The laboratory performed both human and animal mitochondrial DNA testing on the samples, but did not detect either. Although there is no source of nuclear DNA in hair fibers, mitochondrial DNA can be used for identification. The laboratory scientist concluded that the samples were not composed of either human or animal hairs because of the failure to detect any DNA, and determined the samples to be made from synthetic fibers. As a result, the attorneys filed a civil lawsuit on behalf of the woman against the manufacturer of the hair extensions.

The hair extension manufacturer hired a forensic expert in hair identification to examine the evidence to determine whether the material is human hair or some other fiber. Upon microscopical examination, it was immediately clear that this was human hair (Figures 4.2 and 4.3). The shaft of a human hair is composed of a cuticle, a cortex, and a medulla. The cuticle and medullary patterns are used for species identification. In this case, the hairs had an imbricate cuticle scale pattern and an amorphous medulla, with a medullary index less than one-third, thus indicating that they were of human origin. A confirmatory test was performed which consisted of curling the wetted hair, and letting it dry. Hair retains its curls when dried because of

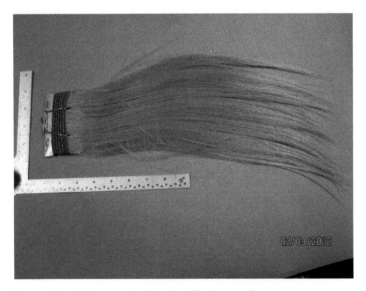

02/09/2012

Figure 4.1   A photograph of the blond hair extension.

Figure 4.2 A brightfield photomicrograph of one of the hairs from the blond hair extension as viewed with a 20-times objective, showing the hair shaft including the cuticle, cortex, and medulla.

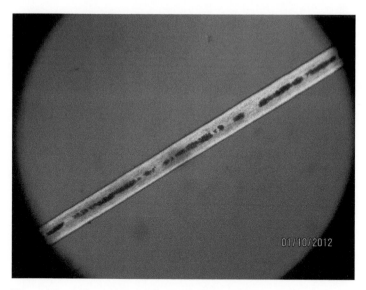

Figure 4.3 A photomicrograph of one of the hairs from the blond hair extension as viewed with a 20X objective between crossed polars with a 530-nm wave plate compensator.

its disulfide bonding intermolecular forces, while synthetic fibers do not. The forensic expert thus concluded that the hair extensions were made from human hairs.

The question still remains about why the molecular biological analysis of the hair extensions did not result in detection of mitochondrial DNA. There are two suspected explanations. First, it is unknown what chemical processing is done to the hairs in preparation for being made into hair extensions, but this could potentially destroy the mitochondria or inhibit its detection. Second, although the private laboratory indicated that they used FBI protocols for the hair analysis, which consisted of DNA extraction and subsequent analysis by quantitative polymerase chain reaction (qPCR), no data was provided verifying their results or showing that proper controls were performed. Regardless if it was the hair extension processing or incompetence of the private laboratory, the conclusion that the hair extensions were not human hair was false.

When the results of the microscopical analysis were presented to the opposing attorney, the case was settled.

This case demonstrates numerous principles, but one in particular is the importance of having the right knowledge to know what questions to ask when investigating a scientific problem. The plaintiff's attorney assumed that DNA analysis would answer the question, and that the laboratory was competent to do the work. However, they did not have sufficient knowledge to frame the question correctly, thus they sought the wrong type of analysis, which resulted in an incorrect conclusion. Microscopical analysis is the ideal method for answering a question regarding hair identification, and ultimately provided the appropriate resolution to this case.

4.2 Blue Yarn Case

John A. Reffner, Ph.D.[1] and Brooke W. Kammrath, Ph.D.[2,3]

[1] John Jay College of Criminal Justice, New York, NY, USA
[2] University of New Haven, West Haven, CT, USA
[3] Henry C. Lee Institute of Forensic Science, West Haven, CT, USA

A young woman was bludgeoned to death in the basement of her home early in 2006. There was a considerable amount of evidence linking her ex-boyfriend to the crime. This included cell-phone records showing him in the vicinity of the crime scene and written threats to the victim. There was also a piece of tape recovered from the cheek of the victim. This was determined by the local forensic laboratory to be from a rare brand of black vinyl tape that was only sold to three companies in the area, one of which was the suspect's place of employment. A roll of this was found in the home of the suspect. Subsequent DNA testing of the tape from the victim's check also showed a Y-chromosome short tandem repeat (Y-STR) profile that was consistent with that of the suspect or someone with the same paternal lineage.

The prosecuting attorney wanted more evidence associating the suspect with the crime, and thus sought help from a private forensic scientist. In particular, a piece of blue yarn was recovered from the floor of the basement next to the victim's body, and no source for its origin was found at the home of the victim. The prosecutor's assumed that the yarn came from the clothing of the perpetrator, and sent detectives to the home and workplace of the suspect to collect blue fabrics. Sadly, the state laboratory no longer performed trace fiber examinations, which is why he needed the assistance of the private forensic scientist.

The forensic scientist was given the unknown blue yarn and one known fabric for comparison which was obtained from the workplace of the suspect. Upon macroscopic and stereomicroscopic inspection of the unknown, it determined to be a light blue yarn with a z-twist composed of natural fibers (Figure 4.4). The known fabric however was composed of dark blue synthetic gear-crimped fibers (Figure 4.5). Thus, it was immediately clear that there was no possibility of their coming from a common source. This concluded the examination.

Although additional analyses (e.g., PLM, infrared microspectroscopy, ultraviolet-visible microspectroscopy) could be performed to further verify this conclusion, the differences in the morphologies of the known and questioned yarns was sufficient for determining that they could not share a common source. In this case, the stereomicroscope was useful for answering a question regarding the probative value of the fiber evidence. Microscopy quickly eliminated the possibility of an association between the known clothing and the evidence fibers, thus adeptly preventing the wasting of both time and resources. Ultimately, the fiber evidence in the case of the blue yarn was a red herring, but other evidence convinced the jury to return a guilty verdict. While the fiber evidence did not help to solve the problem of who committed this murder, it must be remembered that fibers and fabrics have been useful in a plethora of other cases to demonstrate linkages between victim, perpetrator and crime scene (e.g., the Atlanta Child Murders Investigations in Chapter 9, and as Connecticut Murder Case in Chapter 7).

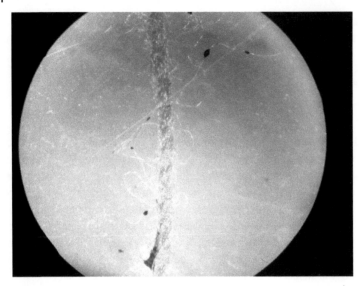

Figure 4.4 The unknown light blue yarn, as viewed with a stereomicroscope. The dark colored material adhering to the yarn was identified as the victim's blood by the state forensic science laboratory.

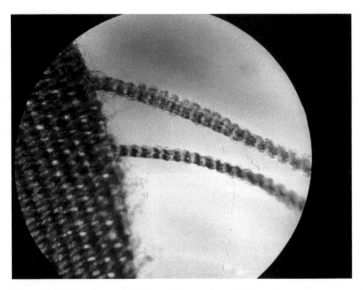

Figure 4.5 The known fabric swatch, as viewed with a stereomicroscope, showing the dark blue synthetic continuous gear-crimped fibers.

4.3 eBay Evidence

John A. Reffner, Ph.D.[1] and Brooke W. Kammrath, Ph.D.[2,3]

[1] *John Jay College of Criminal Justice, New York, NY, USA*
[2] *University of New Haven, West Haven, CT, USA*
[3] *Henry C. Lee Institute of Forensic Science, West Haven, CT, USA*

An elderly woman's home was burglarized, and among other items, a relatively rare book was stolen. A few days later, the woman saw this same book being sold on eBay. She became convinced that this was her book, and purchased it from eBay with the hope it would contain physical evidence to aid in the investigation of the burglary. When the book was received, inside was a black-colored hair. She suspected that her Asian neighbor had committed the crime, and believed this was his hair. The woman took the book and hair to the local police department with the expectation that they would be able to identify the hair as having come from the neighbor, thus proving he was the perpetrator.

The officer was good friends with a Lecturer at the University of New Haven, and asked if there was someone capable of determining the ancestral origin of the questioned hair. The lecturer approached a professor with expertise in microscopy, and she agreed to examine the hair.

Forensic examination of hairs for ancestral origin determination is well documented in the literature (Petraco & Kubic, 2003; Robertson, 1999; Robertson & Brooks, 2021). Ancestral origin determinations are based on hair morphological characteristics that are able to differentiate hairs into three anthropological classifications: originating from European, African, or Asian ancestry, respectively (Figure 4.6). Hairs with a European ancestry exist in a range of colors, show slight variations in their diameters, are flexible, and have an oval cross-sectional shape. Hairs with an African ancestry are usually black in color with heavy pigmentation, show large variations in their diameters, and have a flattened cross-sectional shape. Hairs with an Asian ancestry are usually black in color with a dense pigmentation, show no variations in their diameters, and have a circular cross-sectional shape. Currently, mixed-racial hair is becoming more common which results in head hairs not corresponding to these traditional anthropological classifications. Hairs not exhibiting classical characteristics are labeled as being from an unknown ancestral origin.

The investigator came to the professor's laboratory with the hair, and stayed to observe the examination. This would ensure the chain of custody of the potential evidence item remained intact. Using a stereomicroscope, the unknown hair was observed to be a dark black hair with a slight curvature, measuring approximately 3 inches in length. The microscopist then mounted the hair in a drop of distilled water on a microscope slide, and observed it with a higher magnification brightfield microscope. Immediately, it was obvious that the hair was not of human origin and instead came from an animal. A useful characteristic for differentiating human and animal hairs is the medullary index, which is the ratio of the width of the medulla (a canal that runs through the center of some hairs, and is either filled with air or cells) to that of the entire hair. The value for the medullary index of humans is less than one-third while that for animals is usually greater than one-half. The questioned

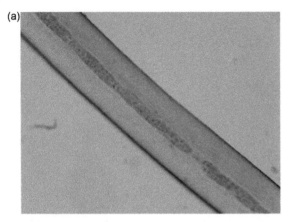

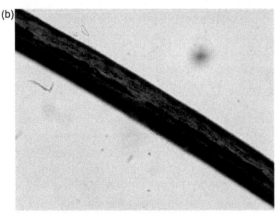

Figure 4.6 Brightfield photomicrographs of human hairs originating from individuals of European (a), African (b), and Asian (c) ancestries, as viewed with a 20-times objective.

hair in this case had a medullary index greater than one-half, which readily proved it to be of animal origin. Additionally, the medulla had a pattern consistent with animal hair.

The entire analysis took a few minutes, and when the professor revealed the results to the investigator, he enthusiastically exclaimed "that's magic." The professor replied "it's not magic, its microscopy!" The professor then had the investigator look through the microscope at several examples of human and animal hairs to educate him in the power of the microscope to reveal these size and shape distinctions. Subsequently, the investigator contacted the burglary victim and revealing that her eBay evidence provided no useful information for resolving her case. However, it provided him with the satisfaction that he had done his due diligence in investigating this case in a timely manner without requiring significant tax payer time and resources.

References

Petraco, N., & Kubic, T. (2003). *Color atlas and manual of microscopy for criminalists, chemists, and conservators*. CRC Press.

Robertson, J. R. (1999). *Forensic examination of hair*. CRC Press.

Robertson, J. R., & Brooks, E. (2021). *A practical guide to the forensic examination of hair: From crime scene to court*. CRC Press.

4.4   An Attractive Contamination

Dale Purcell Ph.D.[1,2,*] *and Paul Schields, Ph.D.*[1]

[1] *Curia Indiana, West Lafayette, IN, USA*
[2] *Chemical Microscopy, LLC., West Lafayette, IN, USA*
[*] *Current affiliation*

A pharmaceutical product was prepared in a stainless-steel reactor. Upon quality assurance testing, the liquid suspension was found to contain a contaminant consisting of microscopic black particles. This is an expensive problem for the pharmaceutical company, not only because the batch needed to be discarded but also because they needed to conduct a scientific investigation into the source of these particles. After examination by internally employed scientists, the company sought consultation with an expert microscopist.

When the microscopist arrived at the pharmaceutical laboratory, a prepared slide of the sample was presented for examination. Upon microscopical inspection, the microscopist observed that the opaque fine particles were aligning themselves in their liquid mounting medium. This indicated that there was a magnetic attraction. He then asked for a magnet, to test this hypothesis. When the magnet was brought in close proximity to the slide, the particles moved toward the magnet, thus validating the initial observation. After reporting the conclusion that the contaminant was magnetic, possibly iron, the pharmaceutical supervisor rejected this possibility stating that all components of the batch processing system were composed of high-grade stainless steel which is nonmagnetic. Although he doubted the veracity of the microscopist's conclusion, the supervisor agreed to examine all processing parts with a magnet after they were thoroughly cleaned. Using the magnet, they were able to locate a magnetic moving part within the system which was recently replaced with the wrong alloy. This magnetic part was then exchanged with the high-grade stainless-steel version, which solved this contamination problem.

4.5   Identifying Metallic Particulates in Pharmaceutical Sample Holders

Dale Purcell, Ph.D.[1,2,*], Stephan X.M. Boerrigter,[1,3,*] and Paul Schields, Ph.D.[1]

[1] Curia Indiana, West Lafayette, IN, USA
[2] Chemical Microscopy, LLC., West Lafayette, IN, USA
[3] Triclinic Labs, Inc., Lafayette, IN, USA
[*] Current affiliation

Many pharmaceutical ingredients have crystalline properties and can be characterized by X-ray diffraction (XRD) of their powders. When performed in transmission mode, the powder XRD sample must be contained in an X-ray-transparent sample holder. Polymer films are commonly used for this purpose where the sample is sandwiched between two of such layers as depicted in the schematic in Figure 4.7.

During manufacturing, the polymer films are perforated in order to facilitate specimen preparation. At some point it was noticed that several batches of the polymer films contained black particulates. Sample contamination is not acceptable and the question was what these particles were composed of and where did they come from? Were they introduced locally or were they introduced during manufacturing?

Initially, various local sources were considered. However, multiple particulates were found between stacks of films in three freshly opened boxes of the same lot. The sample preparation was clearly not where the particulates were introduced and the local laboratory could be eliminated as the source.

A simple notification to the manufacturer, at that time, in order to replace the affected lot would have been an insufficient resolution. The polymer films had been routinely used for a long time and an important question was if these particulates had been present in prior analyses and how may they have affected the results. In order to know how the powder XRD results may have been affected, it was important to identify the composition of the particulates so that previously collected data could be inspected for its signal accordingly.

Furthermore, the question was if it could be guaranteed that any future potential contamination of the polymer films could be prevented? Obviously, it was important to identify the exact nature of the particulates.

For further investigation into the nature of the particulates, the help of the company's microscopy group was sought. Fortunately, the research facility has a well-equipped

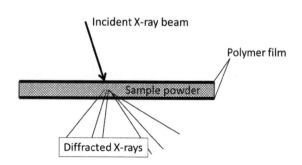

Figure 4.7   Schematic of sample analysis in transmission-mode powder XRD. The layer thicknesses are not drawn to scale.

Incident X-ray beam

Polymer film

Sample powder

Diffracted X-rays

microscopy facility as microcopy offers an important contribution to research on the solid state of pharmaceutical materials. Pharmaceutical microscopy is predominantly used for solid-state form analysis, size and shape determination, and identification of unknown foreign particulates is also a common goal. Often these analyses are applied to drug product, packaging materials, excipients, and raw materials. Foreign particulates are a subgroup of contaminant materials and they might also be introduced in materials used for sample preparation for otherwise routine solid-state form (polymorph) identification using powder XRD as in the present case (http://go.amriglobal.com/WF-2020-701-WP-Identifying-Particulates-01-Landing-Page.html). Interestingly, most foreign particulates are detected during the manufacturing processes or in quality control tests of the product. Nonetheless, however, between Oct 1, 2017, and Jan 4, 2021, 11% (34 of 321) of product recalls resulted from foreign materials and particulate contamination in on-the-market pharmaceuticals (U.S. FDA MedWatch).

Identifying pharmaceutical foreign particulates is essentially equivalent to using microscopy for forensic investigations involving trace evidence which may contain dust from the environment. These identifications require a large and diverse knowledge of manufactured and raw materials. The success of a contaminant investigation almost always requires a collaboration of experts to solve these complex problems. Some of the instruments and techniques used to identify foreign particulates include, among others, stereomicroscopy, PLM, photomicrography, infrared and Raman microspectroscopy, scanning electron microscopy with energy dispersive X-ray spectrometry (SEM-EDX), and powder XRD, notwithstanding classical spot tests, microchemical tests, and a variety of separation techniques. The ultimate goal of the investigation is to correctly identify the foreign particulate, determine the point of entry or root cause of the contamination, and to abate the cause.

The particulates in the current case study appeared dark to the unaided eye and several were removed from the films by rolling them onto a collection paper and by picking. Stereomicroscopy revealed two distinct morphologies: thin flakes and ribbons; both displayed a metallic sheen (Figures 4.8, 4.9, and 4.10). The flakes were

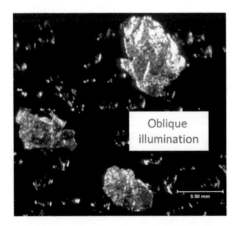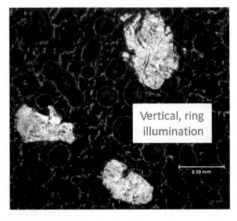

Figure 4.8 Stereomicrographs of magnetic, metallic flakes found on plastic films and mounted on an adhesive carbon, conductive tab on an aluminum SEM pin stub. This same preparation and stub was used for the SEM-EDS results shown in Figure 4.11.

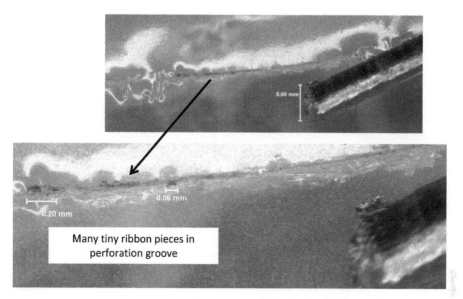

Figure 4.9 Stereomicrographs of metallic, curled ribbons on plastic film near to and in perforation.

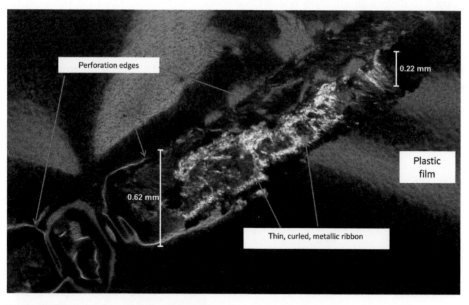

Figure 4.10 Stereomicrographs of a metallic ribbon found embedded in a perforation on a plastic film. The texture, size, and shape of the ribbon and perforation are similar.

observed to be attracted to a magnet, suggesting they were indeed metallic and most likely contained iron.

Metal particulates are a frequently encountered contaminant and their source is typically not immediately obvious. Metal particles can be produced by smelter emissions, wear and tear on gears, needle strike, and cutting tools.

The intimate association and similar morphology of the ribbon and perforation in Figure 4.10 indicates the ribbon was embedded and imprinted by the perforation process.

SEM-EDS analysis was used to determine the elemental composition of the metallic flakes (Figure 4.11). The elemental composition revealed the flakes were composed mostly of iron with small amounts of manganese, carbon, and oxygen.

Figure 4.12 shows an estimated 70 μg of flakes and ribbons mounted on a Si zero-background holder (ZBH). The flakes were magnetically separated from the ribbons for separate powder XRD analysis. Powder XRD analysis of such small amounts of material with a convention X-ray source requires counting times of several hours. Sufficient signal-to-noise was achieved to identify a number of characteristic powder XRD peaks (Figure 4.13). The peaks matched the α form of iron (Fe), specifically ferrite, with a body-centered cubic (bcc) unit-cell length of 2.866 Å. Comparison of the background generated by an empty (blank) ZBH further reveals the presence of iron by its high contribution to the background due to fluorescence (Figure 4.13).

The minor elemental components of carbon and manganese in the metallic flakes and the single, major phase of α Fe are typical of the elemental and phase compositions of plain carbon steel. The absence of detectable chromium rules out stainless steel (Bramfitt & Benscoter). The minor amount of oxygen in the elemental composition from EDS is consistent with the native-oxide surface layer formed on most metals exposed to oxygen in air. EDS is a surface-specific technique. The EDS signal in the present examples was generated by a 20 keV electron beam which samples roughly the top 1.5 μm of the surface.

Powder XRD was also completed on the ribbon particulate. The powder XRD pattern of the ribbons matches face-centered cubic (fcc) aluminum (Al) metal (Figure 4.14). The separation of the ribbons and flakes for separate powder XRD analysis was important for

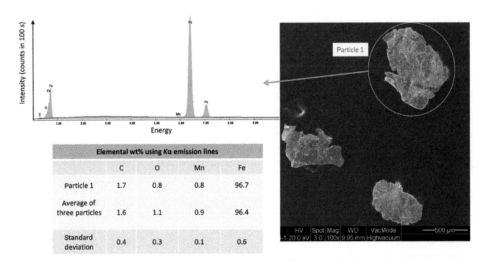

Figure 4.11 SEM-EDS spectrum and image of the flakes shown in Figure 1 with the elemental weight percent (wt%) for particle 1, the average wt% of the three particles, and the standard deviation. The absence of chromium indicates the flakes are not stainless steel ([3] Bramfitt (2022)/ASM International & Benscoter).

Figure 4.12 Stereomicrograph of the flakes and ribbons on a Si ZBH.

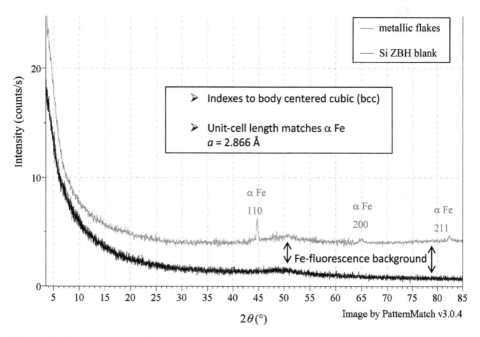

Figure 4.13 The powder XRD pattern of the metallic flakes matches α Fe. The pattern of the blank ZBH shows the contribution to the background radiation caused by the fluorescence of iron in the metallic flakes.

interpretation because each characteristic peak of α Fe is practically coincident with a corresponding characteristic peak for Al. The differences in relative intensity of the peaks in the observed pattern and reference patterns could not be reliably used to detect the presence of each in a mixture. This is because the particulate preparations deviate from a perfect powder preparation which has an ideal, random orientation of all crystals.

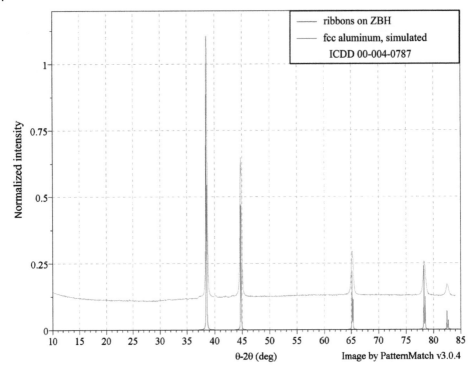

Figure 4.14 The powder XRD pattern displayed by the metallic ribbons matches a reference pattern for fcc aluminum, ICDD 00–004-0787 (International Center for Diffraction Data). The pattern was offset along the vertical axis for clarity.

Therefore, the relative intensity in a pattern of such a mixture would not be reliable enough for identification of α Fe.

The location of the Al ribbons embedded in and next to the perforation groove and their similarity to the size, shape, and texture of the perforation groove indicates they are likely involved with the production of the perforation. Perforation dies and punches are commonly made from stainless steel and commonly have a softer material to stop the blade. Repeated impaction of metals is known to produce metal flakes and ribbons.

In conclusion, considering all information gathered, it was deemed most likely that the particulates were introduced into the polymer films during the perforation step of the manufacturing process.

The scenario and results described above were communicated to the manufacturer of the films. Our detailed information was instrumental in the rapid resolution of the issue. The manufacturer confirmed that, indeed, the contamination must be due to the perforation step. They discovered that the issue was caused by a perforation frame holding the die used for perforating the plastic. This frame is composed of plain, carbon steel. In addition, the stainless-steel perforation punch was too sharp and was impinging the perforation into the Al backing which produced Al ribbons. The punch was subsequently shaped to the appropriate sharpness. Daily cleaning of the die frame was initiated along with installing strong magnets to remediate any Fe particulates which might be present above the plastic films during perforation in the future.

4.5.1 Experimental Details

Stereomicroscopy was performed using a Leica MZ6 stereomicroscope equipped with a Spot Insight color digital camera. Images were acquired at ambient temperature using Spot Advanced software version 4.5.9 build date June 9, 2005. Micrometer bars were inserted onto the images as a reference for size, and the particle sizes were measured using an eyepiece-reticle scale calibrated using a NIST-traceable stage micrometer.

Energy dispersive X-ray spectroscopy was performed using an EDAX™ Sapphire X-ray detector mounted on an FEI Quanta 200 SEM and EDAX™ Genesis v. 3.5 software was used to collect and analyze the spectra. Samples were prepared for analysis by particle picking and were mounted on a carbon-adhesive tab supported on an aluminum SEM stub. The analysis time, recorded in detector live time, was 100 seconds, the amp time was 10 μs, and the beam voltage was 20.0 kV. Both instruments were calibrated using NIST-traceable standards. SEM images were collected using xTm (v. 2.01), and analyzed using XT Docu (v. 3.2). Magnifications reported on the SEM images were calculated upon the initial data acquisition. The scale bar in the lower portion of each image is accurate upon resizing the images and was used to make size determinations. The data-acquisition parameters are displayed in the information bar at the bottom of each image.

Powder XRD patterns were collected with a Panalytical X'Pert PRO MPD diffractometer using Cu radiation produced using a long, fine-focus source and a nickel filter. The diffractometer was configured using the symmetric Bragg–Brentano geometry. Prior to the analysis, a silicon specimen (NIST SRM 640f (National Institute of Standards & Technology Certificate, Standard Reference Material 640f, 2019)) was analyzed to verify the consistency of the Si 111 peak position with the position provided in the NIST certificate. A specimen of the sample was prepared on a silicon ZBH. Antiscatter slits were used to minimize the background generated by air. Soller slits for the incident and diffracted beams were used to minimize broadening from axial divergence. Diffraction patterns were collected using a scanning, position-sensitive detector (X'Celerator) located 240 mm from the specimen and Data Collector software v. 2.2b.

References

http://go.amriglobal.com/WF-2020-701-WP-Identifying-Particulates-01-Landing-Page.html U.S. FDA MedWatch.

Bramfitt, B. L., & Benscoter, A. O.. Metallographer's Guide: Irons and Steels (#06040G), Chapter 1. In *Introduction to Steels and Cast Irons.* © 2002 ASM International®.

International Center for Diffraction Data, 12 Campus Blvd, Newtown Square, PA 19073, USA.

National Institute of Standards & Technology Certificate, Standard Reference Material 640f. (2019, Feb 26). *Line position and Line Shape Standard for Powder Diffraction* (Silicon Powder).

4.6   15ᵗʰ-Century Block Books at The Morgan Library & Museum: The Role of Microscopy in Unraveling Complex Ink Formulations[1]

Federica Pozzi[1,2] and Elena Basso, Ph.D.[1]*

[1] *Department of Scientific Research, The Metropolitan Museum of Art, New York, NY, USA*
[2] *Centro per la Conservazione ed il Restauro dei Beni Culturali "La Venaria Reale", Venaria Reale (Torino), Italy*
[*] *Current affiliation*

The Morgan Library & Museum, New York, holds the largest collection of block books in the western hemisphere. These objects are rare examples of early multipage printed books, produced as woodcuts with blocks carved to include text and illustrations by means of an alternative method to typographic printing (Figure 4.15). Mostly created

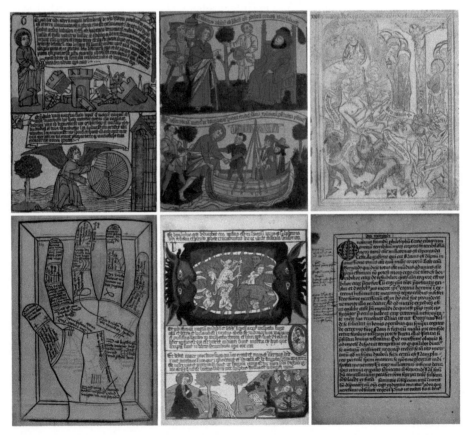

Figure 4.15   Examples of block books analyzed in this study. From top left to bottom right: leaf 37 of *Ars Moriendi* (PML6), leaf 2 of *Apocalypsis Sanctis Johannis* (PML1051), leaf 12b of *Ars Moriendi* (PML3), plate 12 of *Die Kuns Chiromantia* (PML10), leaf 43v of *Apocalypsis Sanctis Johannis* (PML8), and plate 15 of *Ars Moriendi* (PML4).

---

1 Published with kind permission of the *European Physical Journal* (*EPJ*).

in Germany and the Netherlands in a very specific timeframe, between 1460 and 1490, these volumes often had religious content and aimed at a popular audience. Among the books owned by The Morgan are several versions of the *Apocalypsis Sancti Johannis* (Apocalypse of Saint John), *Biblia Pauperum* (Bible of the Poor), *Ars Moriendi* (The Art of Dying), *Ars Memorandi* (The Art of Memory), as well as rare copies of the *Canticum Canticorum* (Song of songs) and *Die Kunst Chiromantia* (The Art of Palm-Reading). Featuring naïve, but powerful images of angels and monsters, saints and sinners, demons and biblical figures, all these titles have precedent in earlier illuminated manuscripts. Because block books were likely printed on demand, each extant copy is unique, as the format, woodblocks, paper stock, printing ink, and pigments were selected when volumes were ordered.

Information on the materials used to create block books is currently limited. No historical recipes of inks or pigments specifically employed for this type of objects have survived to the present day; however, several 15th-century texts addressing the preparation of inks used to stamp woodblocks on leather or textile are extant, which were likely referred to in the manufacture of block books and other contemporary woodcuts along with recipes for manuscript illumination (Needham, 2009; Oltrogge, 2015; Ricciardi & Beers, 2016). For instance, Cennino Cennini describes printing on cloth with black inks based on vine black or lampblack, as well as several other colors such as yellow from saffron, red from brazilwood, red lead or vermilion, green from verdigris, blue from indigo, and white from lead white (Thompson & Cennini, 1954). Other sources mention iron gall writing inks prepared from gallnuts and vitriol, in which the addition of extra gum Arabic had the purpose to yield a more viscous formulation better suited for printing (Stijnman, 2013). Similarly, the coloring of block books was likely achieved by following historical recipes for pigments and dyes to be used in illuminated manuscripts, mentioning brazilwood, folium, lac, vermilion, red lead, mosaic gold, orpiment, lead tin yellow, weld, verdigris, ultramarine blue, azurite, and lead white, among other materials (Cardon, 2007; Merrifield, 1999; Oltrogge, 2015; Ricciardi & Beers, 2016; Thompson, 1956).

While inks used in typographic printing and pigments from illuminated manuscripts have been the subject of a fair body of literature, scant evidence can be found on inks and colors employed specifically in block books. When examined under magnification with a stereomicroscope, most inks in The Morgan's volumes appear rather thin and translucent, ranging in color from a variety of brown shades to gray and black hues (Figure 4.16). While many of the books are printed in a single ink color, a few contain more than one. In terms of polychromy, some pages include only a single accent color, most frequently red, while others feature random areas with multiple hues, and a few are extensively handpainted. Most shades in the bound volumes are still vibrant, while noticeable fading has occurred on loose prints. Occasional inconsistences in the coloring were also observed, likely pointing to later additions and, thus, prompting further investigation.

In this context, in 2019, a collaborative endeavor between The Metropolitan Museum of Art (The Met) and The Morgan Library aimed to explore the materials and techniques used to create block books through careful documentation and scientific analysis of a selection of 13 volumes in The Morgan's holdings. This research project was carried out as part of The Met's Network Initiative for Conservation Science (NICS), a 6-year program aiming to share the museum's resources, expertise, and state-of-the-art scientific research facilities

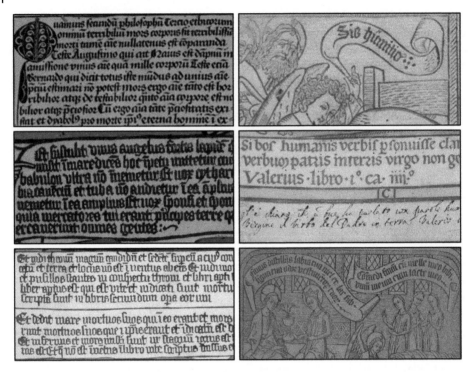

Figure 4.16 Examples of printing inks found in The Morgan's block books. Photomicrographs were taken using a stereomicroscope at low magnification.

with partner institutions in New York City. Initially, examination of the paper and watermarks with beta-radiography, as well as transmitted and raking light, offered clues as to the country of origin and approximate dating of these books. Then, a combination of in-situ, non-invasive scientific analyses with portable and handheld instrumentation and microinvasive investigations of microscopic samples by means of benchtop equipment afforded a detailed characterization of the array of inorganic and organic materials constituting the colors and printing inks found on 29 different pages. The scope of this scientific study was manifold, addressing questions related to provenance and dating of the block books, as well as to the nature and composition of both polychromy and printing inks. While an article about this work was recently published in the *European Physical Journal Plus* (Pozzi et al., 2021), for the purposes of the present chapter a brief summary of the overall findings will be followed by a more focused description of the results concerning the inks.

All watermarks found in the block books examined – over 50 – are compatible with types identified in the Netherlands and Germany in the second half of the 1400s. The concurrent presence of different watermarks in each volume supports a theory according to which block books were printed on demand and one at a time, using paper that was readily available in the printers' workshops. Scientific analysis of the color palette revealed an extensive use of both mineral pigments and natural dyes, often applied in mixtures and with an intent of yielding distinctive color hues and visual effects. Most of the materials detected were in common use in the 15th century for manuscript illumination, albeit fewer occurrences of expensive pigments were detected in this study, in accordance with the popular nature and target audience of block books. In a few cases, modern colors such as emerald green, Prussian blue, chrome yellow, and barium sulfate were also identified. In addition, a detailed examination of the volumes' polychromy exposed the presence of outliers – namely, pigments used in the 15th century, but not frequently found in the present work. These data shed light on the presence of inserted replacement pages, clearly pointing to later interventions. In some instances, moreover, degradation phenomena – including fading – have obscured the full variety of color shades originally applied by painters.

For what concerns printing inks, conservators had initially hypothesized that the different colors observed might correspond to specific compositions: in particular, brown colors were deemed potentially indicative of iron gall inks, while gray and black tones were believed to be related to carbon-based formulations. A combination of non-invasive and micro-invasive techniques was thus employed to corroborate or challenge this theory through materials analysis.

Among the non-invasive tools, infrared reflectance false color (IRRFC) imaging aimed to provide preliminary evidence about the possible use of iron gall inks, typically showing as red (Havermans et al., 2003). However, while as expected the brown inks yielded a fairly intense response, some of the black and gray ones – those presumably based on carbon – also appeared red using this type of imaging technique. Similarly, analysis with portable X-ray fluorescence (XRF) spectroscopy led to ambiguous results. Indeed, in some cases, brown and black inks were respectively characterized by relatively intense iron (Fe) peaks or by the absence thereof; in other instances, however, this proposed correlation did not subsist. For example, a black ink on plate 15 of *Ars Moriendi* (PML4) generated Fe and copper (Cu) peaks, which, nonetheless, were also detected in the paper substrate, suggesting that such ink could be actually based on carbon. On the other hand, in some gray or gray-black inks, including those on leaf 37 of *Ars Moriendi* (PML6), a significantly higher content of Fe and Cu was found in the ink itself compared to the bare paper, along with smaller amounts of manganese (Mn) and zinc (Zn). The combined detection of these four elements is typically indicative of the use of an iron gall ink (Hahn et al., 2004, 2005; Stijnman, 2013). In light of the inconclusiveness of the data collected, a decision was made to remove microscopic samples from a selection of representative locations of the block books in order to determine the inks' composition as accurately as possible with the aid of microscopy and micro-invasive techniques.

Results from this second phase of the analytical campaign revealed, for the printing inks, a more complex formulation than initially assessed, which could only be uncovered thanks to microscopic observations and the use of benchtop instruments equipped with a microscope and camera. Careful visual inspection under magnification, alongside the analysis of samples with micro-Raman and SEM-EDX spectroscopy, enabled us to correlate morphological data with vibrational and elemental information by shedding light on the inks' composition at a microscopic scale. Results showed that mixtures of carbon-based and iron gall inks are quite common both in brown and black inks. For instance, Raman analysis of the brown inks on leaves 48 and 49 of *Apocalypsis Sanctis Johannis* (PML5), for which XRF had detected relatively intense Fe peaks, confirmed the use of iron gall inks, yielding characteristic bands at 571, 1286, 1338, 1479, and 1587 cm^{-1} (Figure 4.17a). However, observation of the specimen under a microscope uncovered the additional presence in both instances of scattered black particles, which gave rise to broad bands at ~1335 and ~1585 cm^{-1} (Figure 4.17a) and were thus identified as carbon. On the other hand, a black ink on leaf 37 of

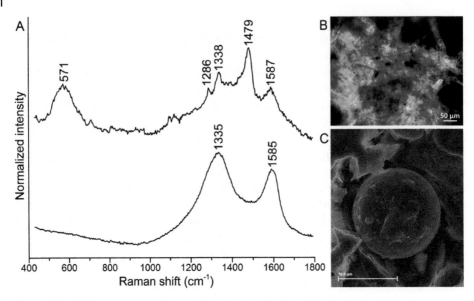

Figure 4.17 (a) Raman spectra of iron gall (top) and carbon-based (bottom) inks obtained upon removal of samples using a benchtop micro-Raman system with 785-nm excitation. (b) Photomicrograph of a sample removed from leaf 64v of *Apocalypsis Sanctis Johannis* (PML7), showing large brown areas of iron gall ink, as well as sporadic carbon-based black and red lead orange particles. (c) Backscattered electron image showing the spherical morphology of a carbon-based particle.

Ars Moriendi (PML6), initially thought to consist of carbon, gave rise to XRF spectra with prominent Fe peaks. Raman spectroscopy, in this case, detected the typical signals of carbon-based inks along with weaker bands assigned to iron gall formulations. Overall, these results are in accordance with medieval recipes for writing inks, in which iron gall inks, carbon-based inks, and mixtures of the two are mostly mentioned (Stijnman, 2013). Moreover, they are also in line with data reported in the literature on the identification of a mixture of carbon-based and iron gall inks in an Apocalypse block book analyzed at the Rijksmuseum, Amsterdam, with XRF and infrared reflectography (Stijnman, 2013).

Remarkably, in some of the block books examined, red mineral pigments such as red lead, vermilion, and iron(III) oxide were found individually or in combination as additional components of the iron gall and carbon-based ink formulations. These may have served the purpose of adjusting the ink's color, thickening its consistency, and/or functioning as a drier to help the ink set up more rapidly. In this case, too, microscopy was crucial as it first uncovered the presence of an abundance of orange particles in the samples under investigation (Figure 4.17b), challenging the initial results of non-invasive analysis. Some of these inks are located, respectively, on leaves 2 and 64v of two versions of the *Apocalypsis Sanctis Johannis* (PML1051 and PML7) – both inserted replacement pages printed with modern pigments.

Among the volumes studied are a few uncolored block books including the *Speculum Humanae Salvationis* (PML20680), created as an interesting mixture of woodblock and type. A combination of microscopic examination and scientific analysis revealed the use of three different ink formulations on its two pages. The light brown ink on leaf 22, on the right of the double folio, was found to be a mixture of carbon-based and iron gall inks; the brown woodblock ink at top left of leaf 21 contains both inks, along with iron(III) oxide; while the black below, on the same page, is pure carbon, whose identification was unambiguously confirmed by morphological data through the detection of micrometric spheres in back-scattered electron images (Figure 4.17c). In addition, SEM-EDX elemental analysis showed an unusually high content of calcium (Ca) and phosphorus (P) in all three inks that, along with the identification of individual particles of calcium phosphate, pointed to the possible addition of bone or ivory black to the ink formulation.

This project has substantially enriched the current knowledge of the materials and techniques employed in block books, linking them to single-leaf woodcuts of the 15th century and placing them within a medieval illumination tradition. In addition to shedding light on fascinating aspects surrounding the fabrication of these rare, unusual objects, this study supported ongoing efforts to digitize all block books in North America and to update the bibliographic descriptions of those owned by The Morgan Library for display and storage purposes.

References

Cardon, D. (2007). *Natural dyes. Sources, tradition, technology and science.* Archetype Publications.

Hahn, O., Kanngießer, B., & Malzer, W. (2005). *Studies in Conservation, 50*(1). https://doi.org/10.1179/sic.2005.50.1.23

Hahn, O., Malzer, W., Kanngießer, B., & Beckhoff, B. (2004). *X-Ray Spectrometry, 33*(4). https://doi.org/10.1002/xrs.677

Havermans, J., Aziz, H. A., & Scholten, H. (2003). *Restaurator, 24*(2). https://doi.org/10.1515/REST.2003.88

Merrifield, M. P. (1999). *Medieval and Renaissance treatises on the arts of painting.* Courier Corporation, Dover.

Needham, P. (2009). *Studies in the History of Art, 75.*

Oltrogge, D. (2015). Chapter 4: Colour stamping in the late fifteenth and sixteenth centuries: technical sources and workshop practice. In A. Stijnman & E. Savage (Eds.), *Printing colour 1400-1700: History, techniques, functions and receptions.* Brill.

Pozzi, F., Basso, E., & Snyder, R. (2021). *European Physical Journal Plus, 134*(4). https://doi.org/10.1140/epjp/s13360-021-01335-w

Ricciardi, P., & Beers, K. B. (2016). Chapter 2: The illuminators' palette. In S. Panayotova, D. Jackson, & P. Ricciardi (Eds.), *Colour: The art and science of illuminated manuscripts.* Harvey Miller Publishers.

Stijnman, A. (2013). In C. Fabian, M. Knoche, M. Linder, E. Mittler, W. Schmitz, & H. Vogeler (Eds.), *Bibliothek und Wissenschaft* (Vol. 46). Harrassowitz Verlag.

Thompson, D. V. (1956). *The materials and techniques of medieval painting.* Courier Corporation, Dover.

Thompson, D. V., & Cennini, C. (1954). *The craftsman's handbook* (pp. 115–117). Courier Corporation, Dover.

4.7 The Critical Value of Microscopy within Pharmaceutical Development

Pauline E. Leary, Ph.D.

Noble, Stevensville, MD, USA

4.7.1 Introduction

The evaluation of pharmaceutically relevant chemicals using light microscopy is an irreplaceable capability. In the early stages of the drug-development process when the amount of sample available for testing is limited, microscopical methods are especially useful. In these instances, an understanding of the physical and optical properties of a sample such as crystal habit, particle size, and polymorphic form of new-chemical-entity (NCE) candidates for development is possible using minute amounts of sample.

The integration of spectroscopic methods such as infrared and Raman with light microscopes extends the value of the microscopical analysis even further, providing spectroscopic information that is used to chemically characterize the NCE and support drug-development efforts. Among other things, these methods are useful to establish or confirm properties such as salt form, polymorphic form, amorphous content, water content, and solvent content from the smallest of samples.

Thermomicroscopy, which integrates the heating and cooling of samples with microscopical analysis, can provide information about the NCE itself, as well as how the NCE behaves when mixed with excipients and other inactive ingredients. The NCE will be in intimate contact with these other ingredients during manufacture and shelf life of the final dosage form. Understanding potential interactions, therefore, can be useful to help predict the best form of the NCE for development. Although historically the likes of Ludwig and Adelheid Kofler, Maria Kuhnert-Brandstätter, and Walter McCrone used heating/cooling stages and interpreted results based solely on melting points and other optical properties they directly observed using the microscope, today's systems seamlessly integrate thermomicroscopy methods with infrared and Raman microspectroscopy, enabling a more complete characterization of physical and chemical interactions.

Ultimately, the goal within pharmaceutical development is usually to synthesize an NCE that will serve as the active pharmaceutical ingredient (API) in a safe and effective final dosage form. While microscopical methods have value throughout this process, it is early in the drug-development life cycle that these methods can be most useful. Not only because they provide insight that will guide development, but also because this type of data can be used to support patent applications and the defense of these patents in Court. In this case study, my goal is to provide you with examples that demonstrate the value these microscopical methods offer to the pharmaceutical scientist, especially those that offer the most value when applied early in the drug-development process.

4.7.2   Crystal Habit and Crystal Structure Using Light Microscopy

Identification of crystal habit using light microscopy is useful, relatively simple to perform, and inexpensive. API crystals are processed to manufacture the tablet or other solid dosage form of the final formulation. The type and amount of processing employed strongly depends on the morphology of the crystallized particles. Further, the impact of shape is particularly relevant on the filterability and compactability of the produced powder, e.g., by affecting its mechanical and chemical stability. Particularly in the pharmaceutical industry, needle-like crystals are highly undesirable products of the crystallization step since they are characterized by very poor flowability and are inclined to the formation of significant amounts of fines when treated (Salvatori & Mazzotti, 2018). Figure 4.18 shows examples of drugs of different crystal habits including squares, rectangles, rosettes, combs, and needles.

Evaluation of crystal-growth behavior and exploitation of crystallization conditions to optimize API crystal habit on a microscopic scale is also important. It can provide insight that can be applied when evaluating potential API manufacturing pathways. It can also be used

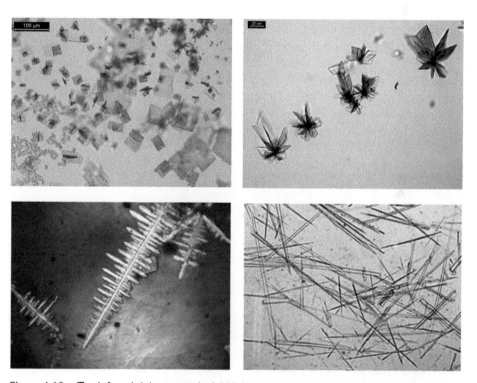

Figure 4.18   (Top left and right, respectively) Light microscope images of yellow-colored rectangular and square plates (benzylpiperazine crystal complexes), and rusty red/orange-to-yellow colored pleochroic rosettes (codeine crystal complexes; Brinsko (2015)/U.S. Department of Justice/Public Domain); (bottom left) comb-like crystals visualized through an infrared attenuated total reflection infrared microscope objective almost at the point of contact with the diamond internal reflection element of the objective, visualized through crossed polarizers with compensation (cocaine crystal complexes); (bottom right) light microscope image of needle-shaped crystals (amoxicillin from urine sediment; Poloni (2020)/Renal Fellow Network).

to identify new crystal habits. Figure 4.19 shows a photomicrograph of an API recrystallized from melt when visualized through crossed polarizers with full-wave compensation. In this image, two different crystal habits of the drug are observed. Needle-like crystals growing from spherulites comprise most of the field of view. There is also a single hexagon-shaped crystal highlighted by a yellow arrow in the top left area of the sample. Prior to this preparation, this hexagon-shaped crystal habit for this API had not previously been observed. These crystallization experiments were performed to understand thermal behavior exhibited by this API during differential scanning calorimetry analysis, but the identification of this new crystal habit, which is a more manufacture-friendly crystal habit, occurred. Subsequent evaluation of crystal structure using optical microscopy of these hexagon-shaped crystals as well as the needles growing from spherulites showed that although these crystals exhibited two different crystal habits, they were both of the same crystal structure. Crystal structure can be a very important for both drug stability and development, as well as for legal reasons such as intellectual property (IP) rights protection.

4.7.3 Polymorphic Identification Using Thermal Microscopy

As mentioned previously, the crystal structure (or polymorphic form) of an API is important within the pharmaceutical industry. From a legal perspective, polymorphic form may be important for IP-rights considerations. From a drug-development perspective, polymorphs are important because they can exhibit differences from each other including dissolution, bioavailability, stability, and potency.

Thermomicroscopy, or the integration of a heating and cooling stage with a microscope, can be useful to evaluate polymorphic form. In its simplest form, identification of melting

Figure 4.19 API recrystallized from melt into two different crystal habits (needles growing from spherulites and a hexagon-shaped crystal) visualized through crossed polarizers using full-wave compensation.

point can be used to identify and differentiate polymorphs when they exhibit different melting temperatures. Figure 4.20 and 4.21 show thermomicroscopy results from two different polymorphic forms of an API. Note that simply by heating the sample and identifying melting temperatures, polymorphs can be easily detected.

In other instances where the melting points of polymorphs are the same, more elaborate heating and cooling experiments may be designed to better interrogate the sample. The results of one such example is shown in Figure 5. These images were taken during the reheat of an API recrystallized from melt and visualized using a polarized light microscope with crossed polarizers. During the initial heat of the API as received from the manufacturer, which is not shown in this figure, only a single melting temperature of the API was observed. However, after the initial melt was cooled and the API recrystallized from melt was reheated (shown in Figure 4.22), the API exhibited two additional melting temperatures. Interestingly, the three melting temperatures, i.e., the melting temperature of the API as received from the manufacturer and the two melting temperatures of the recrystallized API, were all different from each other. This indicated that this API exhibited at least three different polymorphic forms.

It is important when performing this type of evaluation to verify that the API does not decompose with melt. Evaluation of melt with decomposition can be performed using differential scanning calorimetry and/or thermogravimetric analysis, both techniques routinely used within the pharmaceutical industry for this purpose. Decomposition with melt is common for small molecules like pharmaceutical APIs and must be considered during observations performed during thermomicroscopy experiments.

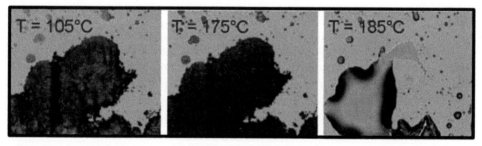

Figure 4.20 Thermomicroscopy of polymorph I of an API (see Figure 4.21 for analysis of polymorph II of this API).

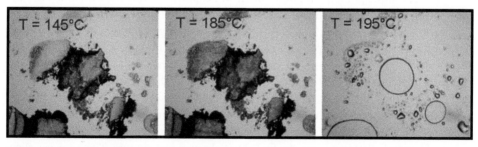

Figure 4.21 Thermomicroscopy of polymorph II of an API (see Figure 4.20 for analysis of polymorph I of this API).

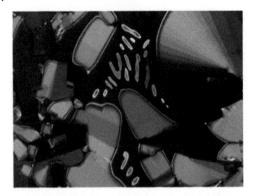 

Figure 4.22 Photomicrographs of an API during heat. Prior to collection of these photomicrographs, the API as received from the manufacturer was melted and then recrystallized from melt on a microscope slide. These images are from the second melt of the API and collected using a polarized light microscope through crossed polarizers. The left image shows the melting of the one polymorphic form of the drug; note that in the right image, this polymorphic form is completed melted while the second polymorphic form remains in the solid phase.

Interestingly, subsequent attempts to isolate the initial polymorphic form of this API (which exhibited the highest melting temperature) were not successful. However, upon storage at elevated temperature, the recrystallized API converted back to the polymorphic form of the API as received from the manufacturer.

4.7.4 Identification of Solid-Solid Transitions Using Thermomicroscopy

It is also possible to directly observe solid-solid transitions using thermomicroscopy. Such transitions from one solid-state form to another are important. They may indicate an event such as a conversion from an unstable polymorphic form to a more stable one, or the release of water from a hydrated form of the API. Hydrated forms are important because differences in molecular weight due to hydration level can impact final dosage-form strength. An understanding of these events is critical from a drug-development perspective to insure the desired solid-state form of API is present in the solid dosage form.

Figure 4.23 shows photomicrographs of an API captured during the heat of the sample through its melt. These images are collected through slightly crossed polarizers. Note that during the heat between 25 °C and 125 °C, the sample gets darker due to the way the light is scattered by the sample as it is heated. At 135 °C, though, it is obvious that the solid state of the drug is converting from one form to another. By 145 °C, the conversion is complete. By the time the temperature reaches 155 °C, this second form is completely melted.

Isothermal holds performed during thermal microscopy can also be used to evaluate solid-solid transitions. Figure 4.24 shows the same field of view collected at three different time points of a sample that was isothermally held at elevated temperature. These images show how the crystal front of the more stable solid-state form (Form II) moves across the field of

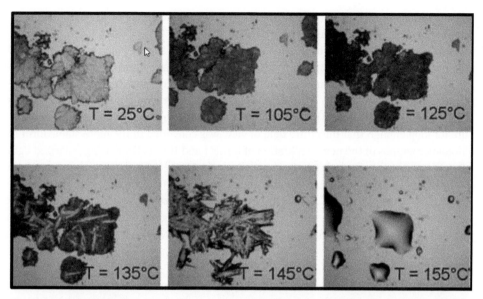

Figure 4.23 Thermal microscopy of an API showing both a solid-solid phase transition and melt as the sample is heated.

Figure 4.24 Isothermal hold resulting in a transformation of one solid-state form of an API to another observed using thermal microscopy.

view as the less stable form (Form I) of the sample is converted from the less stable form to the more stable form. This type of behavior is very important to understand so that the manufacturing pathways can be optimized, and the final dosage form of the drug product is stable.

4.7.5 Identification of Solid-Solid Transitions Using Thermomicrospectroscopy

The ability to visualize the transformation of a sample during thermomicroscopy, as has been demonstrated, has value. Observations made during these types of evaluations can help to not only identify the best solid-state form an API for development, but also can help to determine the best process for manufacture of the final dosage form. While direct observations are important and offer a great deal of insight about the sample and its behavior, the ability to add spectral information to this type of analysis can provide very useful information about chemical changes the sample is undergoing that can be directly compared with visual observations.

As an example, after the thermomicroscopy analysis of the API shown in Figure 4.24 was performed, it became important to understand the exact nature of the transition occurring. For this reason, thermomicroscopy with Fourier transform-infrared (FT-IR) microspectroscopy was performed. For this analysis, a new microscopic preparation of the less stable form of the API (form I) was prepared and held isothermally at an elevated temperature for long enough that the entire field of view was converted to the more stable form II. Photomicrographs of the new preparation of forms I and II visualized using polarized light microscopy through crossed polarizers are shown in Figure 4.25. FT-IR spectra from both forms of the API were collected and compared using an FT-IR microspectrometer. These spectra are shown in Figure 4.26. The infrared spectra of these two polymorphs were almost identical. The only difference was a slightly different absorption at the carbonyl absorption band. The similarity in the infrared spectra of polymorphs is expected since polymorphic

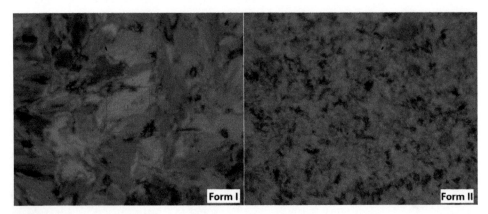

Figure 4.25 Photomicrograph of two polymorphs of the same API that is shown in Figure 7. (left) Less stable solid-state form (form I); (right) more stable solid-state form (form II).

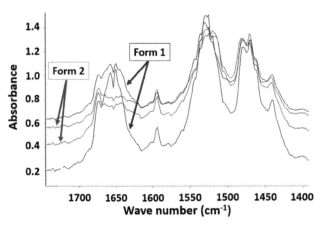

Figure 4.26 Infrared spectra from two polymorphs of the API shown in Figures 4.24 and 4.25. Form I is the less stable polymorph; form II is the more stable polymorph.

forms are chemically identical; the only difference between them is the structural arrangement of the atoms. In this example, differences in the structural arrangement were shown (based upon the infrared spectral data) to be due on some level to the arrangement of the carbonyl bond. This information enabled a better understanding of the nature of the sample and how the sample was converted from one form to another over time.

4.7.6 Evaluation of Component Interactions Using Thermal Microspectroscopy

It is also important to understand how different components of a formulation interact with each other. These interactions can have a direct impact on product performance including drug efficacy, dissolution, toxicity, and stability.

Many microscopists through the years have studied small molecules using a microscope, and scientists like Maria Kuhnert-Brandstätter applied microscopy methods, including the use of thermomicroscopy, extensively to pharmaceuticals (Kuhnert-Brandstätter, 1971). These microscopists relied upon careful evaluation of optical properties and melting-point values to establish the type of interactions between different chemicals. Today's microscopists have the added benefit of adding infrared and Raman spectral data with these thermomicroscopy methods so that not only can the melting and cooling behavior of the API be directly observed but a spectroscopic evaluation of these interactions is also possible.

Fusion methods are useful to evaluate the interactions between different chemicals. Fusion methods include the methods and procedures useful in research and analysis which involve heating a compound or a mixture of compounds on a microscope slide and observations made on the heating of the preparation, on the melt itself, during solidification of the melt, and during further cooling (McCrone, 1957). One approach is to melt a sample of a lower melting temperature on a microscope alongside a chemical of a higher melting temperature. As the chemical with the lower melting temperature melts, it interacts with the chemical with the higher melting temperature. In an area referred to as the "zone of mixing", which is the region of the microscope preparation where to two chemicals react, the nature of their interaction can be observed. The interactions are monitored, and the effect each compound has upon the other is evaluated.

The simplest interaction resulting from this type of fusion preparation is formation of a eutectic. A eutectic is a uniform fine-grain physical mixture of the two components. The eutectic has a single melting temperature that is always lower than the melting temperature of either starting material. The eutectic is a physical mixture of the two components, so the infrared spectrum of the eutectic is a mixture of the infrared spectrum of each component. There should be no additional absorption bands present in the eutectic spectrum that are not present in the spectrum of each starting material, and only minimal shifts in bands resulting from chemical interactions between the two components. If the infrared spectrum shows additional or different absorption bands or significant shifts in band position, it is likely that the interaction between the two components is chemical in nature, such as when an addition compound is formed (Reffner et al., 2008). In more interesting interactions, an addition compound may be formed along the zone of mixing, and then eutectics between the addition compound and each of the starting chemicals may be observed.

Figure 4.27 shows a fusion preparation of two pharmaceutically relevant chemicals that formed a simple eutectic along the zone of mixing. The image on the left of the figure shows the preparation below the melting temperature of the eutectic. The image in the middle of the figure shows the preparation at the melting temperature of the eutectic. The image on the right shows the preparation at a temperature above the melting temperature of the eutectic but below the melting point of either starting compound. Figure 4.28 shows a composition diagram of a two-component mixture that forms an addition compound along the zone of mixing, as well as a eutectic along each side of the starting chemicals between the newly formed addition compound and each starting chemical. Although understanding the nature of these interactions is very important from both drug stability and performance perspectives, they also are important for IP rights considerations.

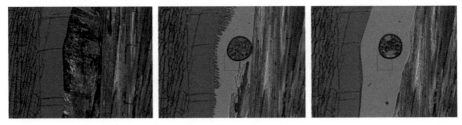

Figure 4.27 Photomicrographs of a simple eutectic during its melt, visualized through slightly crossed polarizers. (left) Below the melting temperature of the eutectic, (middle) at the melting temperature of the eutectic, and (right) above the melting temperature of the eutectic, but below the melting temperature of either starting chemical.

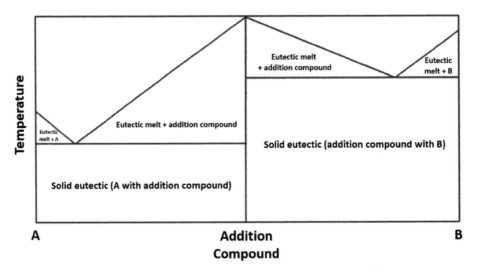

Figure 4.28 Composition diagram of two-component system forming an additional compound with a eutectic on each side of the addition compound.

4.7.7 Conclusion

The use of microscopy within the pharmaceutical industry is important because it provides the ability to visually observe a microscopic sample at the earliest stages of the drug-development process. Even this relatively simple task can help to inform the drug-development process since properties such as crystal habit can be critically important during manufacture. The integration of the microscope with accessories such as heating and cooling stages, as well as infrared and Raman detectors has extended the value of the microscope to incredible limits by enabling thorough physical and chemical evaluations of the sample under environmental conditions that mimic manufacturing procedures.

References

Brinsko, K. M., Golemis, D., King, M. B., Laughlin, G. J. and Sparenga, S. B. (2015). *A modern Compendium of Microcrystal Tests for Illicit Drugs and Diverted Pharmaceuticals.* McCrone Research Institute. Chicago: United States Department of Justice, 2015-2016.

Kuhnert-Brandstätter, M. (1971). *Thermomicroscopy in the Analysis of Pharmaceuticals (International Series of Monographs in Analytical Chemistry).* Pergamon Press. 1971. 978-0080069906.

McCrone, W. C., Jr. (1957). *Fusion Methods in Chemical Microscopy.* Interscience Publishers, Inc. 1957.

Poloni, J. A. T. (2020). *Urine Sediment of the Month: Drugs & Crystalluria.* RenalFellow. org. [Online] May 28, 2020. Cited March 14, 2022, from https://www.renalfellow. org/2020/05/28/urine-sediment-of-the-month-drugs-crystalluria

Reffner, J. A., Leary, P. E., & Giordano, J. L. (2008). The imaging advantage of FT-IR spectroscopy. *Spectroscopy.* Special Issues, August 1, 2008.

Salvatori, F., & Mazzotti, M. (2018). Manipulation of particle morphology by crystallization, milling, and heating cycles: experimental characterization. *Industrial & Engineering Chemistry Research, 57*(45), 15522–15533. https://doi.org/10.1021/acs.iecr.8b03349

5

Rely on the Fundamentals

Introduction

With the introduction of multivariate statistics and advanced instrumentation, there is a tendency to skip the fundamentals and go straight to a more sophisticated analysis. This is a mistake. We must always rely on the fundamentals. As defined by *Oxford Dictionary*, the fundamentals are "a central or primary rule or principle on which something is based" (https://search.yahoo.com/search?p=fundamentals&fr=yfp-t&fp=1&toggle=1&cop=mss&ei=UTF-8). There are general fundamental principles for performing good science as well as specific to individual fields of study and instruments. General chemistry, biology, and physics courses are required because they provide fundamental knowledge about matter, life, and energy. These are the foundation for all further studies. With any instrument, there are basic fundamentals that must be applied to obtain reliable and reproducible data. For example, proper alignment and quality sample preparation are examples of important fundamentals for using both light and electron microscopes.

There is an anecdote which aptly demonstrates the problems with failing to apply the fundamentals. A rancher needs to tell the difference between two horses. So, he counts the number of legs and teeth, measures the horse's height and the length of the tail and mane, and tracks how fast they run. Ultimately, with these numbers, he concludes that the black horse is in fact different from the white one. The absurdity of this is obvious. As previously mentioned, observation is an integral part of the scientific method, and the first step in the process (see Introduction and Chapter 2). In this anecdote, the observation of color would be considered subjective, that does not make it less valid or useful for differentiating the horses. The same should be applied to all observations, including those examined using a microscope. As discussed in Chapter 4, visual observations are validated by their testing like any other method of scientific analysis.

This chapter provides five case examples where the microscopist's reliance on fundamental scientific principles was essential to answering the question and resolving the problem.

Solving Problems with Microscopy: Real-life Examples in Forensic, Life and Chemical Sciences, First Edition. Edited by John A. Reffner and Brooke W. Kammrath.

5.1 A Mouse, a Soft Drink Can ... and a Felony

S. Frank Platek, MS, FMSA (Ret.) and Nicola Ranieri, BS

US Food and Drug Administration, Office of Regulatory Affairs, Office of Regulatory Science, Forensic Chemistry Center, USA

In 1993, what began as a possible misunderstanding and national reporting of an insulin syringe discovered in a soft drink can, initiated hundreds of consumer complaints of similar incidents reported across the United States. This caused a confidence issue for consumers with a disastrous economic impact to several commercial drink manufacturers (Halverson, 1993). Intentionally tampering or false reporting of product tampering are felonies outlined in the Federal Anti-Tampering Act of 1983 (Fines Enhancement Statutes of 1985; Federal Anti-Tampering Act 1983). The Food and Drug Administration's (FDA) Forensic Chemistry Center (FCC) analyzed samples associated with 235 separate cases in 1993, all involving claims of foreign objects found in beverage cans and bottles. Foreign objects included a wide variety of syringes (insulin, intramuscular, veterinary and equipment lubrication), as well as needles, pins, nails, screws, thumb tacks, safety pins, pieces of glass and metal shards, and even bullets and cartridge casings. Also included in the suspect cases were dead rodents. In the case presented here, stereomicroscopy, polarized light microscopy (PLM), computer-assisted image analysis (IA) and scanning electron microscopy with energy dispersive X-ray spectrometry (SEM-EDX) were used to provide evidence supporting a felony charge of false reporting of a product tampering (Platek et al., 1996, 1997).

In July of 1993, the FCC received information of a consumer complaint alleging an attorney in Somers, New York, had found a dead mouse in a can of diet soft drink. The attorney reportedly purchased the drink from a vending machine in June of 1993 on her drive home and, once opened, took a few drinks along the way. She reported the product tasted "vile" but that it was not her regular soft drink. Upon arriving home, she opened the car door and poured the remaining liquid onto her driveway which then revealed the pink foot of a mouse protruding from the mouth opening of the now empty can. She placed the can with the carcass inside in a plastic bag and stored it in her freezer as she considered it a "novelty" find. She further reported becoming ill with nausea and vomiting for several days.

As the national news continued to report numerous additional cases of foreign objects in containers of soft drinks, fruit drinks, and beer, the consumer decided to contact the soft drink manufacturer. The company made the initial collection of the can that contained a mouse carcass from the consumer. The company shipped the can with the carcass to a private veterinary hospital where the can was cut open and the carcass removed. The carcass, without the suspect can, was shipped to an out-of-state veterinary pathology laboratory where a necropsy and pathological examinations of the tissues removed were performed. The original veterinary hospital returned the empty can to the manufacturer.

The consumer began to pursue financial compensation from the soft drink manufacturer. The manufacturer then contacted the FDA's Office of Criminal Investigations (OCI) for assistance and transferred custody of the suspect can as evidence to an OCI special agent. The agent interviewed the consumer to get an account of the incident. The consumer was informed of the Federal Anti-Tampering Act statutes and potential penalties for false report of product tampering and agreed to sign an affidavit. The suspect can was then sent to the FCC for forensic examinations.

Examination of the can was initially performed using stereomicroscopy. The can body, lid, lift tab, center rivet, and mouth opening tab were examined and photo documented as received. The can was received irregularly cut ~270° around the can body and ~2–3 cm down from the lid rim (Figure 5.1). The can lid had the mouth opening tab bent inward toward the inside of the can with the lift tab intact and held in place by the center rivet (Figure 5.2).

Numerous white colored filaments were observed caught in burrs and jagged regions on the lift tab (Figure 5.3). The white colored filaments were removed and examined using PLM and further compared to a mouse hair reference standard (Figure 5.4). Based upon

Figure 5.1 Suspect soft drink can as received. Scale = mm.

Figure 5.2 Top/lid of suspect soft drink can as received.

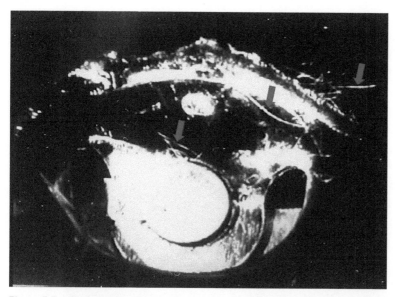

Figure 5.3 Stereomicroscope image of ragged and abraded edge of suspect can lift tab. Note white filaments (at arrows).

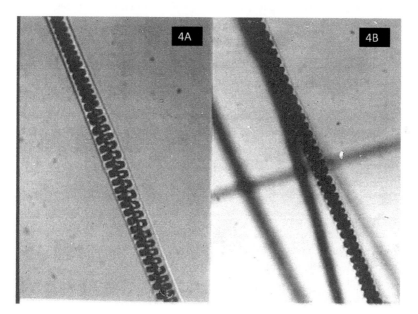

Figure 5.4 PLM comparison of white filaments. Suspect filament (4A) and reference standard mouse hairs (4B).

the size of the filaments, the continuous "laddered appearance" medulla and comparison to a reference standard, the filaments were determined to be mouse hairs.

The contact edge of the lift tab also showed an irregular, gnarled edge with burrs and visible parallel striations resembling abrasions or scratches (Figure 5.5). When compared to a control can lift tab, there was a notable region of metal missing from the edge of the

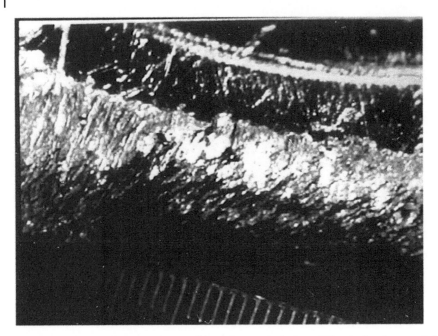

Figure 5.5 Stereomicroscope image of irregular, gnarled edge of lift tab showing parallel striations and missing metal.

suspect lift tab. Image analysis of the edge of the suspect lift tab to the same edge of a control can determined that ~7.1 mm^2 of metal was missing from the suspect lift tab.

Approximately two weeks later, the mouse carcass was received at the FCC as evidence from the veterinary pathology laboratory, post-necropsy and fixed (preserved) in formalin. All thoracic organs (lungs, heart, thymus, etc.) and some of the abdominal cavity organs (kidneys, liver, and adrenal glands) had been removed and were not included with the evidence. Remaining *in situ* and intact was the gastrointestinal tract including the stomach, small intestines, caecum, and most of the large intestine.

Examination of the carcass included stereomicroscopical analysis of the muzzle and the mouth region as well as all dentition. The two upper incisors were tightly abutted (nearly fused), immobile and chisel shaped with a slight indention between the two teeth and the outer corner of the right incisor appeared to be slightly chipped or worn away (Figure 5.6A). Characteristic of most rodents, the lower incisors were independent and slightly movable as the mandibles are not tightly fused at the front of the jaw. The two lower incisors had conical tips and a gray colored staining which was visible on both incisors but was not observed on the other teeth (Figure 5.6C).

The mouth opening in the suspect can lid was observed using stereomicroscopy to have unique marks around the opening on both the top (outer) lid surface and the bottom (inner) lid surface. The marks on the top lid surface appeared to be paired, flat scrapes and cuts in the metal (Figure 5.6B) whereas the marks on the bottom lid surface were separated conical indentions (Figure 5.6D). Using IA, the distance between the tips of the lower incisors was determined to be ~0.3 mm and was consistent with the distance between the paired conical indentions on the inner lid around the mouth opening.

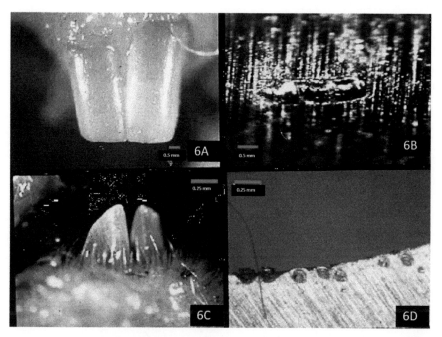

Figure 5.6 6A Upper incisors of suspect mouse showing cutting edge. 6B Bite mark of upper incisors on the outer can lid adjacent to mouth opening. 6C Lower incisors of suspect mouse showing conical tips. 6D Bite mark of lower incisors on the inner can lid adjacent to mouth opening.

Morphometric measurements of the mouse carcass showed it could have passed easily through the mouth opening of the suspect can. The suspected "bite marks" on the suspect can's mouth opening rim would indicate the mouse would have had to be inside the can to make the upper incisor marks on the upper/outer surface and the lower incisor bite marks on the bottom/inside of the suspect can lid. The upward rolling of the edge and burrs of the lift tab edge would also support that the mouse was chewing from inside the can as the rodent would anchor its bite with the upper incisors and chisel with the lower incisors. This further supported a potential source of the gray coloration from the aluminum lift tab on those two teeth. The right lower incisor with the gray coloration was removed from the right mandible and examined by SEM-EDX. The gray colored regions produced a major elemental peak for aluminum where adjacent tooth regions without the gray coloration did not. Using SEM-EDX, the elemental composition of the lift tab produced a major elemental peak for aluminum.

Additional stereomicroscopical examination of the mouth, muzzle, and fur of the carcass further revealed numerous small, elongated, silver-colored particles, all measuring less than 0.5 mm in length caught in the hair (Figure 5.7). Further examination of the lid and bottom inner corner of the suspect can also found a few of the same type of small, elongated, silver-colored particles. Based upon the evidence thus far it was believed the mouse had been in the suspect can and had been chewing on the can's lift tab, which could have been reached from inside the can, as well as the rim of the lid's mouth opening. Assuming the mouse had done this, it was very likely the mouse had ingested fragments of the can's lift tab in chewing the aluminum lift tab. As previously noted, the gastrointestinal tract had not been removed at the original necropsy. The stomach was opened at the FCC and was found to contain a mass of stomach contents presumed to be chewed food and also included numerous small, elongated, silver-colored particles, all measuring less than 0.5 mm in length (Figure 5.8). These particles were visibly consistent to the particles recovered from the mouse's muzzle as well as the suspect can rim and bottom.

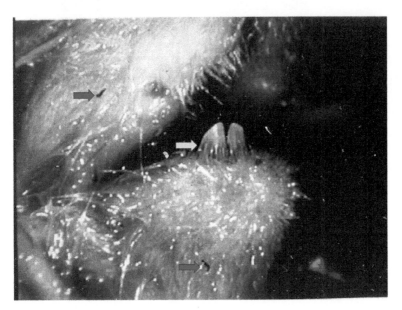

Figure 5.7 Muzzle and lower incisors of suspect mouse showing gray coloration (yellow arrow) and elongated, silver-colored particles in the fur (red arrows).

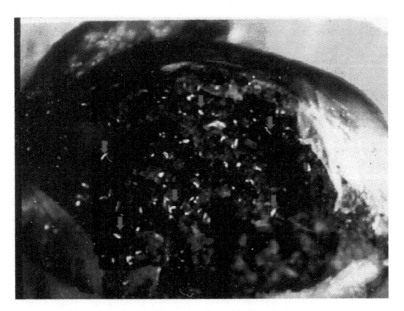

Figure 5.8 Stomach of suspect mouse with a portion of the stomach wall removed showing contents including numerous elongated, silver-colored particles (red arrows).

The small, elongated, silver-colored particles collected from the suspect can and the stomach and muzzle of the mouse were further analyzed and compared using SEM-EDX. Morphometrically, the particles from all locations were consistent to one another in size, shape, and the presence of parallel striations (Figure 5.9). Using SEM-EDX, the particles from all locations appeared nearly identical to each other. Representative particles from the stomach (Figure 5.9A and EDX spectrum Figure 5.9C) and the suspect can's lift tab region (Figure 5.9B and EDX spectrum Figure 5.9D) are shown. All particles were found to produce a single major elemental peak for aluminum and a minor-to-trace elemental peak for magnesium. Other minor-to-trace elemental peaks for sodium, silicon, phosphorus, sulfur, chlorine, potassium, and calcium were attributed to surface contamination of the respective particles dependent upon collection site.

The overwhelming physical and chemical evidence demonstrated the mouse had gnawed on the lift tab and mouth opening of the suspect can while INSIDE the can. It would have been completely impossible for the mouse to have been canned in the soft drink and still be able to chew on any part of the exterior of the can lid or lift tab. It is believed the mouse had entered an opened can (probably empty and discarded) and could

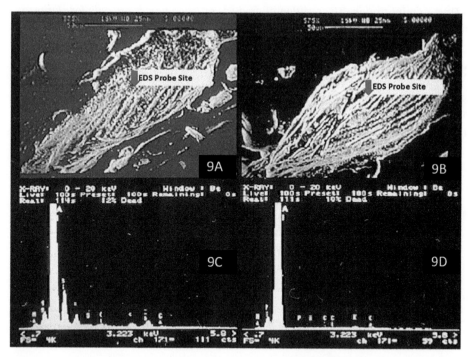

Figure 5.9 (Scales = 50 μm). 9A/C – SEM photomicrograph of elongated, silver-colored particle from the stomach and the resultant EDS spectrum from the indicated probe site. 9B/D – SEM photomicrograph of elongated, silver-colored particle from the lift tab region of the suspect can and the resultant EDS spectrum from the indicated probe site. Both spectra show a major elemental peak for aluminum and minor peak for magnesium. Other minor to trace elemental peaks for sodium, silicon, phosphorus, sulfur, chlorine, potassium, and calcium were attributed to surface contamination of the respective particles dependent upon collection site.

not get out and was subsequently found by the consumer. Since the can's production code showed the can had been filled and sealed three weeks prior to the alleged incident, the relatively good condition of the mouse carcass with no visible necrosis further supports the mouse could not have been canned in the soft drink product as claimed. The defendant in this case was found guilty of falsely reporting a product tampering, extortion, and attempting to flee jurisdiction and was subsequently sentenced to 42 months in a federal prison.

Of the 235 cases examined by the FDA's Forensic Chemistry Center during the 1993 soft drink tamperings, only one foreign object, a stainless-steel machine screw from a production line, was found to be canned in a soft drink product during production.

This case emphasized the use of stereomicroscopy, PLM, SEM-EDX, and computer assisted IA to conclusively refute a consumer's sworn claim that a mouse was sealed inside a sealed soft drink can. It further illustrates the extent of forensic analysis and microscopy resources that can be used to investigate product tampering and protect the American consumer and producers of legitimate products.

Disclaimer

The comments are the authors alone, and do not reflect the opinion or policy of the US Food & Drug Administration. The mention of products or trade names are only for clarity, and should not be considered an endorsement.

References

1983 Federal Anti-Tampering Act, Title 18, USC Section 1365.

Fines Enhancement Statutes of 1985, Title 18, USC Section 3571.

Halverson, G. (1993, June 18). Pepsi reels from reports of product tampering. *Staff writer of The Christian Science Monitor*, New York: The Christian Science Monitor.

Platek, S. F., Ranieri, N., & Wolnik, K. A. (1996, February 22). The mouse that roared - A case of reported false product tampering. *48th Annual Meeting of the American Academy of Forensic Sciences*, Nashville, TN.

Platek, S. F., Ranieri, N., & Wolnik, K. A. (1997, November). A false report of product tampering involving a rodent and soft drink can: Light microscopy, image analysis and scanning electron microscopy/energy dispersive X-ray analysis. *Journal of Forensic Sciences*, *42*(6), 1169–1173.

5.2 Goodrich Corrosion Problem

John A. Reffner, Ph.D.[1] *and Brooke W. Kammrath, Ph.D.*[2,3]

[1] John Jay College of Criminal Justice, New York, NY, USA
[2] University of New Haven, West Haven, CT, USA
[3] Henry C. Lee Institute of Forensic Science, West Haven, CT, USA

Self-sealing, rubber fuel tanks were first patented by the US Rubber Company in 1940 and were widely used in military aircrafts during World War II. All major rubber companies worked together to provide these self-sealing fuel tanks to the war efforts. These tanks consist of several layers of rubbers and fiber reinforcement, creating a chamber for the fuel, and metal fixtures that are imbedded into the rubber. These metal fixtures are imbedded in the fuel tank's rubber prior to the rubber being cured. The steel fixtures are necessary for loading of fuel into the fuel tank and distributing fuel to the aircraft's engines.

One day, at B. F. Goodrich in Akron, Ohio, the manager in charge of quality control for the manufacturing of self-sealing fuel tanks came rushing into the "Works Technical Microscopy Laboratory." He was carrying a bundle of steel fixtures from various rubber fuel tanks that appeared to be rusted. The manager exclaimed "we have a major problem in successfully manufacturing our self-sealing fuel tanks." The problem was that the bond between the metal fixtures and the rubber was failing and fuel was leaking. On visual inspection, the failed metal fixtures were reported as exhibiting signs of corrosion. The microscopist was tasked with solving this corrosion problem; otherwise the production of self-sealing fuel tanks would need to be halted. The metal part was composed of a low carbon steel plated with cadmium. The manager indicated that the cadmium plating was added after earlier studies discovered that humidity and temperature increased the rate of corrosion. Controlling the humidity, temperature and adding the cadmium plating improved the manufacturing process of the self-sealing tanks, but it did not eliminate the failures due to the corrosion problem.

Fortuitously, the microscopist tasked with solving the corrosion problem was a recently hired chemistry student from Akron University. He told the manager that he had seen other samples of these metal fixtures before. In these previous examinations, analysis involved looking at the samples with a stereomicroscope and just reporting that there was evidence of corrosion. Both the manager and the microscopist agreed that just reporting corrosion was not sufficient; understanding the cause of the corrosion was essential to solving the problem. To find the cause required more information. The manager then provided the additional observations that using cells of different rubbers (neoprene, nitrile, butyl, and natural, government rubber styrene or GRS) resulted in different failure rates. Neoprene rubber cells were free of corrosion failures, nitrile rubber had a small rate of failure. Butyl, natural, and GRS rubbers all had excessive corrosion failures. In the prepared mind of this microscopist, these failure rates correlated with the level of sulfur used in curing these rubbers. When he examined the failed metal fixtures using the reflected light stereomicroscope and metallurgical microscope, he observed a yellow material consistent with cadmium sulfide. To test this hypothesis, unused cadmium-plated steel fixtures were exposed to hydrogen sulfide vapors. Surprisingly, no reaction was observed. Upon further testing, the exposure of the cadmium-plated steel parts to hydrogen sulfide vapors was completed in a warm moist environment. This rapidly produced a corrosion product microscopically consistent to that found in the failed rubber fuel tanks. The corrosion was a yellow crystalline material with a hexagonal morphology which is characteristic of cadmium sulfide.

Armed with this new information, the failure problem was solved by excluding the sulfur from contacting the cadmium surface. This correction was easily implemented, and production of the self-sealing fuel tanks was able to continue.

5.3   The Perils of Forgetting the Fundamentals in Criminal Cases

John A. Reffner, Ph.D.[1] and Brooke W. Kammrath, Ph.D.[2,3]

[1] John Jay College of Criminal Justice, New York, NY, USA
[2] University of New Haven, West Haven, CT, USA
[3] Henry C. Lee Institute of Forensic Science, West Haven, CT, USA

A woman was accused of shooting her husband in the head three times with his own firearm while he slept. The victim's Smith & Wesson 9-mm pistol was found at the scene; however, there were no fingerprints recovered on the weapon. It was later proven to be the murder weapon based on microscopical examination of the recovered bullets by a firearms examiner at the local forensic science laboratory. To account for the lack of fingerprints, the police investigator hypothesized that the defendant had put the gun into their dishwasher to remove evidence of her handling the weapon. The defendant maintained her innocence, claiming someone else had broken into their home and committed the crime. Upon examination of the gun, rust and a white residue were observed. Forensic scientists were asked to identify this residue and determine whether its source was consistent with the dishwashing detergent used in the home.

The forensic laboratory's scientists analyzed the residue with X-ray fluorescence (XRF) spectroscopy and X-ray diffraction (XRD), and identified the material as sodium chloride, or table salt. Sodium chloride is a common chemical found in a wide of variety of products, thus this identification alone did not answer the question as to the source of the residue. Consulting criminalists were hired by the defense to further interrogate the results. Upon examination of the XRF and XRD data, trace amounts of aluminum nitride and barium sulfate were identified. This information was then compared to the chemical components of Cascade dishwashing detergent. Cascade dishwashing detergent recovered from the scene had an ingredient list of sodium phosphate, sodium carbonate, chlorine bleach, sodium silicate, and sodium sulfate. It was concluded by the consulting criminalists that the residue could not have come from the dishwashing detergent because the chemistry did not match. The failure to detect phosphates in the residue is of primary importance, as it could not just disappear from the mixture.

The District Attorney was not satisfied with the findings of the initial forensic scientists, and persisted to hire another outside scientist with expertise in scanning electron microscopy. This third expert analyzed the residue with an SEM-EDX. Figure 5.10 shows the data from his analysis.

When the defense's consulting criminalists examined the results of the SEM-EDX analysis, gross incompetence was immediately recognized. In the first EDX spectrum (Figure 5.10A), the elemental data shows the sample contained 5.02% technetium. Technetium is the element with the lowest atomic number (43) in the periodic table that has no stable isotopes, and every form of it is radioactive. Nearly all technetium is produced synthetically, and only minute amounts are found in nature. In the second EDX spectrum (Figure 5.10B), the sample was analyzed and shown to contain 1.56% promethium. Promethium does not exist naturally on earth, and all of its isotopes are radioactive. If this examiner was knowledgeable, detecting these radioactive elements would have raised a red flag about the accuracy of the spectral data. It is impossible for

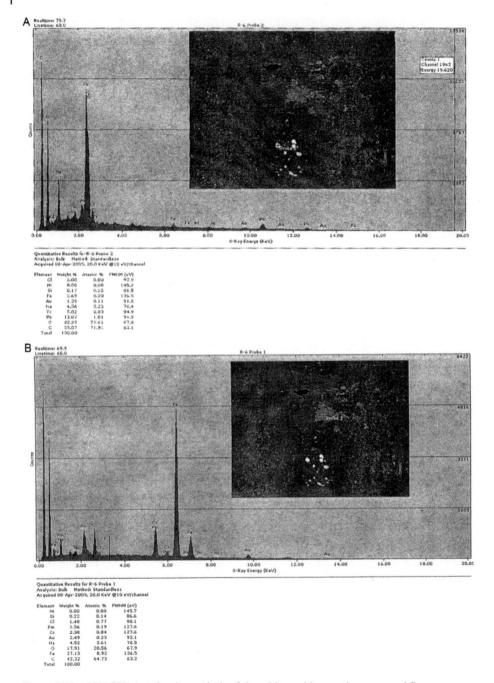

Figure 5.10 SEM-EDX data for the analysis of the white residue on the recovered firearm.

these elements to be present in the sample, especially at these relatively high concentrations, so there is something wrong with the data. This is fundamental science, which he clearly did not understand.

The judge erroneously qualified this scientist as an expert and allowed him to testify in this case, despite the evidence demonstrating him to be unqualified. Sadly, this is not uncommon. The defense challenged the expertise of this scientist, but this motion was denied. Further, the expert testified that the Cascade detergent could have been the source of the residue. When asked about the failure to detect phosphates on cross-examination, he stated that phosphates are flammable and volatile, so they would have evaporated. This is false. Elemental phosphorous is flammable and volatile, not phosphates. Phosphates have different chemical properties than elemental phosphorous – a fact that is taught in every high school chemistry class. So again, he demonstrated an extreme lack of fundamental knowledge of chemistry.

The criminal justice system in this case failed to provide adequate protection against the introduction of bad science. Here, the failure of the expert to recognize the fundamentals of chemistry and SEM-EDX led to the inclusion of bad science in the court. Bad experts make for bad justice.

Physical evidence is only one part of the story. Even if the jury believed that the defendant did not put the gun into the dishwasher, there are other explanations for the failure to recover fingerprints from the firearm. Additionally, other non-scientific evidence was presented by the prosecutor to adjudicate justice, including having the three children wear earplugs while they slept that night. Ultimately the woman was convicted of this homicide.

5.4 Super Bowl White Powder Attacks 2014

John A. Reffner, Ph.D.[1] and Brooke W. Kammrath, Ph.D.[2,3]

[1] John Jay College of Criminal Justice, New York, NY, USA
[2] University of New Haven, West Haven, CT, USA
[3] Henry C. Lee Institute of Forensic Science, West Haven, CT, USA

White powder hoax bioterrorist attacks began in the 1990s, but they only received widespread attention after the 2001 Amerithrax attacks (FBI, 2016). One week after the 9/11 terrorist attacks, letters containing weaponized anthrax spores were sent to politicians and members of the news media, ultimately killing 5 people and infecting 17 others. Following these events, there was a rise in hoax anthrax attacks, which created fear, caused chaos, and damaged infrastructure. These hoax white powder attacks contained a white or off-white powder that consisted of a range of readily available materials. Some commonly used white powders used in these hoaxes include artificial sweeteners, cornstarch, baby formula, and anything one can find in their homes. Although their frequency has diminished since 2001, hundreds of these white powder attacks still occur every year throughout the world.

In 2014, Super Bowl XLVIII was being held in MetLife Stadium in New Jersey. A few days prior to the game, seven nearby hotels received threat letters containing suspicious white powders. First responders rapidly determined that these letters did not contain any hazardous materials, and identified the powders as cornstarch. The media widely praised the rapid response and identification of the hoax terrorist material.

The quick identification of these hoax powders as cornstarch was only made possible by understanding the fundamental science of microscopy and the information it is able to provide. Common protocol for unknown white powder analysis is to examine the sample using PLM after it is known to not contain a bioterrorist agent. The analysis consists of mounting a small amount of the unknown sample onto a microscope slide, adding an oil medium, and examining it with polarized light. PLM is a powerful tool for material identification, with the identification of starches being a prime example. Using brightfield microscopy, starch particles can barely be seen and may only appear to be colorless misshapen ovals or circles. However, when they are examined with cross-polarized light, the particles transmit color and have a characteristic Maltese cross pattern (Figure 5.11). The various sources of starches can be differentiated by their microscopic morphologies, with cornstarch having angular geometries (Figure 5.11A), rice starch being smaller in size and generally circular in shape (Figure 5.11B), and potato starch having the largest size and most varied shapes (Figure 5.11C). The difference in the magnified images of these starches provide a rapid means for their identification.

Knowing the fundamentals of microscopy, and specifically how PLM can be used for the identification of materials, led to the quick resolution of this case. Rapid response,

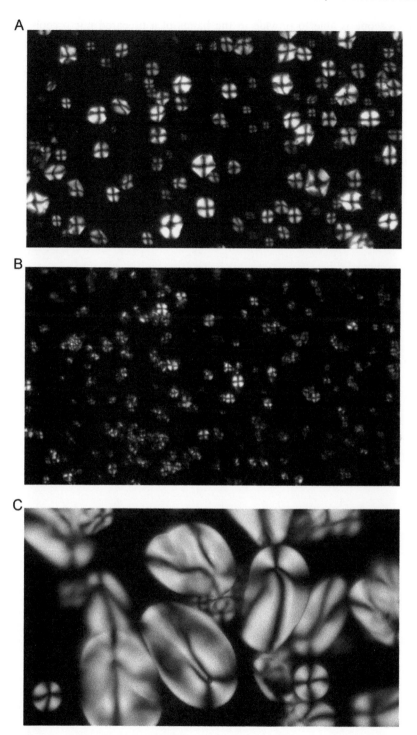

Figure 5.11 PLM photomicrographs (as viewed with a 40-times objective) of cornstarch (A), rice starch (B), and potato starch (C) in crossed polars.

which the microscope is able to provide in the hands of a prepared microscopist, is essential in white powder attack cases to minimize losses and disruptions to life and business.

Reference

FBI. (2016, May 17). *Amerithrax or anthrax investigation.* FBI. Retrieved September 5, 2022, from https://www.fbi.gov/history/famous-cases/amerithrax-or-anthrax-investigation

5.5   College Drug Party

John A. Reffner, Ph.D.[1] and Brooke W. Kammrath, Ph.D.[2,3]

[1] John Jay College of Criminal Justice, New York, NY, USA
[2] University of New Haven, West Haven, CT, USA
[3] Henry C. Lee Institute of Forensic Science, West Haven, CT, USA

College campuses across the United States were ripe with social and political activism during the late 1960s and early 1970s, most notably stemming from America's involvement in the Vietnam War, but also due to issues involving civil rights, gender equality, and apartheid. These student rebellions created significant issues for law enforcement. This case presents an example of a simple problem that arose due to this unrest.

In the late 1960s, the New Haven police department received an anonymous tip regarding a drug-filled party being thrown by a group of Yale students. When police arrived at the scene, there were indications of drug use all over the party. Most notably, there were lines or "rails" of white powder found on tables, which is characteristic of the illicit use of cocaine. Samples of this were documented and collected, and brought back to the police station.

At that time, the field of forensic science was quite different from today. In Connecticut, the State Police had a Bureau of Identification which focused primarily on the pattern evidences: firearms and toolmarks, fingerprints, and handwriting analysis. Illicit drugs were a major societal problem, and suspected drugs were sent to the health department in Hartford for chemical identification. A lack of communication between police officers and scientists in the health department prompted the police to establish their own chemical analysis group. This new unit consisted of two police officers with no formal science education, thus they often sought help from scientists and microscopists at the University of Connecticut's Institute of Material Science.

Coincidentally, the day after the Yale party, a *Leica* salesman was at the Police Academy to demonstrate their new polarized light microscope for use in the new chemical analysis unit. In addition to the two police scientists, a microscopist from the Institute of Material Science was invited to attend this demonstration. A ranking police officer joined the microscope demonstration shortly after it began. As soon as he arrived, he stated, "If that microscope could tell the difference between cornstarch and cocaine, I'd buy that microscope on the spot!" Both the Leica salesman and the microscopist were surprised by this officer's statement due to it being a simple problem to solve using a polarized light microscope. As discussed earlier in this chapter in the case of the "Super Bowl White Powder Attacks 2014", cornstarch is easily identified using a polarized light microscope by its distinct morphology and the appearance of maltese crosses when viewed using crossed polars (Figure 5.11A). Thus, the *Leica* salesman and the microscopist requested additional information from the officer. The ranking officer explained that the Yale students arrested the night before now claimed that it was all a hoax – they had intentionally provoked their own arrest by simulating cocaine with cornstarch in an effort to trick the police. Thus, with his statement about the microscope, the officer was looking for a rapid resolution to this potential false arrest case. The *Leica* salesman fortuitously had a sample of cornstarch in his demo kit, and was able to show the ranking officer how to solve his problem using this polarized light microscope. The three officers were impressed with the ability of the microscope to rapidly, non-destructively, and conclusively solve this problem. Consequently, the microscope was purchased by the police department for use in their chemical analysis unit.

With regard to the case, subsequent polarized light microscopical analysis of the samples recovered from the Yale party identified the white powders as cornstarch. All charges against the students were dropped. Good fortune combined with knowledge of the fundamentals of polarized light microscopy contributed to the rapid solution for this identification problem and timely resolution of this case.

6

Fortune Favors the Prepared Mind

Introduction

In his December 7th, 1854, lecture marking his formal inauguration to the Faculty of Letters of Douai and the Faculty of Sciences at the University of Lille, Louis Pasteur famously declared "*Dans les champs de l'observation le hasard ne favorise que les esprits prepares.*" Translated into English, this stated "In the fields of observation chance favours only prepared minds." There are several variants to this translation, with one of the most common being the title of this chapter: "Fortune Favors the Prepared Mind." (L. Pasteur, *"Discours prononcé à Douai, le 7 décembre 1854, à l'occasion de l'installation solennelle de la Faculté des lettres de Douai et de la Faculté des sciences de Lille"* (Speech delivered at Douai on December 7, 1854, on the occasion of his formal inauguration to the Faculty of Letters of Douai and the Faculty of Sciences of Lille), reprinted in: Pasteur Vallery-Radot, ed., *Oeuvres de Pasteur* (Paris, France: Masson and Co., 1939), vol. 7, p.131.)

Much like a birdwatcher can recognize hundreds of birds from their appearance or call, a pet lover can identify all the different breeds of cats, dogs, and rabbits, and a gardener can name every species of flower and plant in a garden, an expert microscopist can identify a vast range of microscopic particles in the micro world. Preparation of the mind is through education, training, and experience. A microscopist readies his or her mind by preparing and observing a large number of diverse items for interrogation with a variety of different microscopes. A necessary component of this preparation is having the curiosity to investigate all aspects of the natural world, with a focus on asking meaningful questions and attempting to solve challenging problems.

The prepared mind sees things that otherwise may be missed or ignored. The prepared mind of a microscopist enables him or her to recognize informative and relevant details in a magnified image which provide value for answering questions. This chapter includes four case examples which showcase the ability of a microscopist's prepared mind for solving problems.

Solving Problems with Microscopy: Real-life Examples in Forensic, Life and Chemical Sciences, First Edition. Edited by John A. Reffner and Brooke W. Kammrath.

6.1 The NASA Problem

John A. Reffner, Ph.D.[1] and Brooke W. Kammrath, Ph.D.[2,3]

[1] John Jay College of Criminal Justice, New York, NY, USA
[2] University of New Haven, West Haven, CT, USA
[3] Henry C. Lee Institute of Forensic Science, West Haven, CT, USA

The value of the prepared mind was widely advocated by Dr. Walter C. McCrone. Dr. McCrone was an advocate of chemical microscopy who recognized the advantages of having an experienced microscopist explore the micro world to create a vast mental particle atlas which may be used to solve problems through the microscopical examination of materials. His work process for solving contamination problems is detailed in an article published in *American Laboratory*, and "starts with one microscopist who has looked at (and remembers) the microscopical appearance of a few thousand substances. As assurance of success, additional microscopists enrich the databank by having looked at additional substances" (McCrone, 2000, p. 20). This is a proven method of success which has been used in a range of industries, because images are real data, too (Chapter 3). This NASA problem case exemplifies the use of the microscopists' prepared mind for solving problems.

In the early summer of 1969, Dr. McCrone was contacted by a NASA client regarding an urgent problem with the astronaut space suits. Microscopic rodlike crystals were forming in the circulating fluid which had the potential to clog filters and disrupt the flow of fluids in the suits. As this was only a few weeks prior to the Apollo 11 launch, this problem demanded immediate attention. At 11:00 p.m. that same evening, investigators came to McCrone's Chicago laboratory to deliver samples with an expectation that an answer would be reported by 9:00 a.m. the next morning.

Dr. McCrone assembled his team of expert microscopists, which included his wife Lucy McCrone, Don Graber, and one of this book's authors (Dr. John A. Reffner). The sample arrived mounted on a membrane filter, and was examined at 50-times magnification with a polarized light stereomicroscope (Figure 6.1). Walter was the first to examine the specimen,

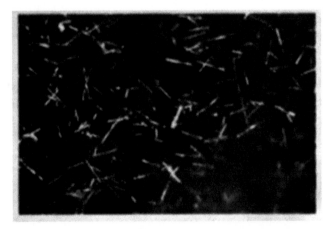

Figure 6.1 Crystalline birefringent rods of the dimer of 2-mercapto-benzothiazole, viewed with crossed polarized light at 80-times magnification. McCrone, 2000/The McCrone Group, Inc.

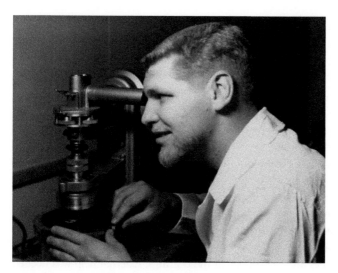

Figure 6.2 John Reffner using the McCrone Associates IR microspectometer in 1963. Perkin Elmer developed this instrument, Model 80, in 1959.

and noted that they were "birefringent rods." Lucy was next, and commented that the sample was colorless with plane polarized light and highly birefringent. Don observed that the samples had parallel extinction and were likely orthorhombic. Last, John examined the sample and within a minute had the answer. He announced that the material was a dimer of 2-mercapto-benzothiazole. This is a chemical substance used in the rubber industry, and thus Dr. Reffner recognized it based on his years of work at B.F. Goodridge in Akron, Ohio. The dimer of 2-mercapto-benzothiazole forms crystals under oxidizing conditions. The monomer is highly soluble; the dimer is not. Although Dr. McCrone was confident in John's microscopic identification, the NASA representative wanted further confirmation. Subsequent tests performed that night included measurements of refractive index and melting point, X-ray powder diffraction and IR microspectroscopy (Figure 6.2). The latter two tests had been developed to perform single-particle analysis by McCrone Associates (Westmont, IL), and thus only one crystal was used for each analysis. All methods verified the initial microscopic identification. The NASA representative confirmed that the 2-mercapto-benzothiazole was a component of the circulating fluid of their space suits, and commented that they had never before seen the dimer. With this material identification, NASA was able to solve the problem with their astronaut suits.

Reference

McCrone, W. C. (2000). Contamination analysis. *American Laboratory (Fairfield)*, *32*(8), 17–24.

6.2 A Train Engine Contamination

John A. Reffner, Ph.D.[1] and Brooke W. Kammrath, Ph.D.[2,3]

[1] John Jay College of Criminal Justice, New York, NY, USA
[2] University of New Haven, West Haven, CT, USA
[3] Henry C. Lee Institute of Forensic Science, West Haven, CT, USA

In 1962, a train broke down near Cedar Rapids, Iowa. Routine inspection of the engine's gauges indicated that the engine was both overheating and that its oil pressure was low. Further mechanical testing by the Kelty Radiator Company (Cedar Rapids, Iowa) showed that the oil cooling tubes were blocked, and thus they needed to be cleaned. After a mechanic cleaned the engine and oil cooling tubes using an abrasive compound via sandblasting, new problems emerged which were determined to be due to damage to the engine bearings. The engine bearings were failing at a faster-than normal rate, and a black carbonaceous residue was observed on the inside of the cooling tubes and the surface of the bearings. External help was needed for solving this problem, and it was sent to Dr. Fred Kisslinger, a metallurgical engineering professor at the Illinois Institute of Technology. Dr. Kisslinger recognized that this was not an engineering problem, and hired a microscopist working for Polytechnic, Inc. in Chicago, IL, to make a study of various samples of residues, abrasive compounds, and the bearings.

The purpose of the study was to characterize and identify the abrasive particles and to determine their presence in the various samples submitted. These samples included 4 residues from different cooling tubes, 3 bearings, and 3 different abrasive compounds. The known abrasives were a silicon carbide abrasive, an aluminum oxide No. 36 abrasive, and an unknown abrasive from a stock pile found under the grit blast cabinet at Kelty Radiator Company.

There is a plethora of different abrasives available for shaping or finishing a range of materials for a wide variety of domestic, industrial, and technological applications. The rubbing action of the abrasive wears away the surface of a material to get a desired outcome, be it for polishing, grinding, buffing, sanding, etc. Abrasive materials are commonly naturally occurring minerals such as sand, corundum, emery, pumice, iron oxide, diamond, and many more. Abrasive materials must be carefully chosen to meet the specifications of the application, as the final properties are dependent on both the size of the mineral and its hardness. Choosing the wrong abrasive can lead to catastrophic results, which was the situation in this case.

The microscopist, who had a deep knowledge of material science, recognized that the first step in this analysis was to identify the mineral component of the unknown abrasive from the Kelty Radiator company. Using optical microscopy and crystallography, via a steromicroscope and PLM, the unknown abrasive was identified as containing an aluminum oxide mineral, specifically corundum (Figure 6.3).

Next, the microscopist examined the residues from the cooling tubes. The black carbon on these residues masked all particles present, thus the samples had to be prepared for analysis of the mineral component. By igniting the residue with a hot flame for 5–10 minutes, the carbon was burned off and then the samples were examined microscopically. In 2 of the 4 residue samples, aluminum oxide abrasive grit was found to be present (Figure 6.4). The minerals in the residue were morphologically similar to those from the unknown abrasive

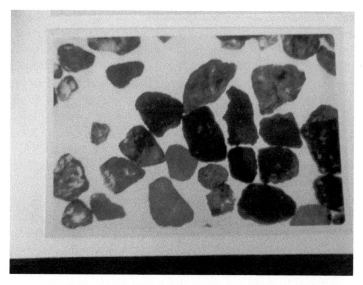

Figure 6.3 Photomicrograph (32 times with 3.2-times ultrapak objective, 10-times ocular, transmitted light) of the unknown abrasive from the stock pile under the grit blast cabinet at Kelty Radiator company.

Figure 6.4 Photomicrograph (32 times with 3.2-times ultrapak objective, 10-times ocular, transmitted light) of abrasive particles removed from one of the residue samples from the cooling tubes.

from the Kelty Radiator company, and were identified as corundum, a standard aluminum oxide mineral.

Last, the bearings themselves were examined using reflected light (or epi) stereomicroscopy with a Leitz ultrapak objective. This specialty objective, which came in a range of magnifications, uses vertical darkfield illumination. The use of darkfield illumination for this examination was necessary to see the surface of the bearings. Brightfield epi illumination would have too much glare to enable a detailed analysis of the reflective surface, however darkfield allowed this detection. But for the microscopist's awareness and understanding of the different illumination and contrast methods, microscopical analysis of the surface of the bearings would not have provided useful and interpretable images. Fortunately, the microscopist's laboratory was equipped with these special objectives. As a result of the use of darkfield epi illumination, aluminum oxide abrasive particles were observed embedded into the surface of the 3 bearings (Figure 6.5). These abrasive particles were the same type as those found in the cooling tube residues, which demonstrated the connection between the damage to the bearing surfaces and the prior cleaning of the engine and oil cooling tubes.

As a result of this microscopical analysis using specialty objectives, the cause of the prematurely failing bearings was identified. The problem was that the engineers were using the wrong abrasive grit to clean their engine and fuel cooling tubes, which left a residue of aluminum oxide particles to damage the bearings. The solution to this was obvious: more attention must be made to the type of abrasive grit used in the cleaning of the train's components, with specific care taken to avoid the use of those containing aluminum oxide particles.

Figure 6.5 Photomicrograph (32 times with 3.2-times ultrapak objective, 10-times ocular, reflected light) of one of the bearings. Note the amber and black glassy particles of abrasive embedded in the metal surface.

6.3 Mianus River Bridge Collapse: Why Do You Need a Microscope to Determine Why a Bridge Fell Down?

John A. Reffner, Ph.D.[1] *and Brooke W. Kammrath, Ph.D.*[2,3]

[1] John Jay College of Criminal Justice, New York, NY, USA
[2] University of New Haven, West Haven, CT, USA
[3] Henry C. Lee Institute of Forensic Science, West Haven, CT, USA

The collapse of the Mianus River Bridge on June 28th, 1983 was one of the most influential highway accidents in history (Figures 6.6 and 6.7). As a tractor-semitrailer truck was traveling on Interstate 95 at 1:30 a.m. through the Cos Cob neighborhood of Greenwich Connecticut, a 100-ft section of the bridge fell 70 ft into the Mianus River. Two tractor-semitrailers and 2 automobiles plummeted through the massive gap into the riverbed, resulting in the death of 3 individuals and serious injuries to 3 others. Due to the time of the event, there were fewer casualties than could have occurred had it been during working hours. Approximately 90,000 vehicles crossed this bridge daily. After medical assistance was provided to survivors and the bodies of the 3 victims were recovered, the focus of the investigation immediately shifted to determining the cause of the bridge collapse.

The Mianus River Bridge was designed in 1955 and its construction was completed 3 years later on December 27, 1958. It was composed of 6 longitudinal sections: 4 suspended spans and 2 anchored spans (one at each end which is connected to the ground). The

Figure 6.6 A 100-foot-long section of the Interstate 95 bridge spanning the Mianus River in Greenwich, CT, collapsed June 29, 1983. Photograph by Bob Child/AP Images – © 2012 The Associated Press. Accessed December 30, 2020, from https://connecticuthistory.org/mianus-river-bridge-collapses-today-in-history

Figure 6.7 Mianus River Bridge Collapse June 27, 1983 – Night of the bridge collapse. Photo by Tom Ryan, https://www.greenwichtime.com/local/article/Then-and-now-I-95-s-Mianus-River-Bridge-11249964.php#photo-13155628, last accessed December 10, 2022.

expansion joints of the bridge, which hold the bridge together when there is movement due to thermal expansion, vibrations, or other ground activities (e.g., seismic or settlement), use a pin and hanger assembly to couple the girders together (Figures 6.8 and 6.9). A gap between two spans of the bridge (the anchored span and the suspended span) is connected via a steel hanger on either side. A pair of steel rods runs the width the bridge and support the spans from underneath while enabling longitudinal movement. These rods are connected

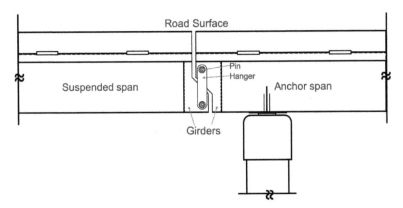

Figure 6.8 Pin and hanger assembly. The suspended span is free to move laterally with thermal expansion and contractions, vibrations, or other ground activities.

Figure 6.9 Photograph of a pin and hanger assembly on the Holley Road Bridge, in Brockport, New York. HistoricBridges.org / https://historicbridges.org/newyork/brockportholley/pinhanger4.jpg last accessed December 10, 2022.

to the hanger through pins, which are steel cylinders with bolts through their centers. Large steel flanges or plates are used as retainers to secure the hangers to the pins. These assemblies are massive in size, with the steel hangers used in this bridge measuring 6 feet 6 inches in height, 18 inches in width, and 2 inches in thickness (Figure 6.10). Due to I-95 being a 4-lane highway, the Mianus River bridge was composed of two separate but parallel pin and hanger assemblies – one for each direction of travel.

On the morning of the 28th, one of the authors of this book, Dr. John Reffner, received a phone call from the Director of the Connecticut Forensic Science Laboratory, Dr. Henry C. Lee, regarding this tragic event. Dr. Lee inquired whether he was available to examine the physical evidence of the bridge collapse. At this point in time, the cause of the bridge failure was unknown and there were questions regarding the involvement of criminal activity (e.g., explosives or sabotage). Parts of the bridge that had fallen into the riverbed were recovered by scene investigators and sent to the forensic laboratory in Meriden, Connecticut for analysis to determine a possible cause of the bridge failure. Dr. Reffner drove to the forensic laboratory immediately to begin a physical evidence examination.

It was fortuitous that Dr. Reffner was the scientist called to assist with this forensic investigation. While the Assistant Director of the Institute of Material Science at the University of Connecticut, he gained meaningful experience with failure analysis. He often collaborated with metallurgical professors to study fracture surfaces using primarily stereomicroscopy and scanning electron microscopy. This experience provided Dr. Reffner with knowledge of the different types of fractures and how microscopy is valuable for their differentiation.

There were several recovered parts of the bridge at the forensic laboratory, with two items having particular value. First, an approximately 3-inch portion of the 1-inch diameter steel

Figure 6.10 Photograph of the recovery of the hanger that fell from the broken pin and caused the collapse of the Mianus River bridge. Photo by Mel Greer/https://www.greenwichtime.com/local/article/Then-and-now-I-95-s-Mianus-River-Bridge-11249964.php#photo-13155622/last accessed December 10, 2022.

bolt that was part of the pin assembly showed considerable deformation. This portion of the bolt was threaded and still had the nut on one end, thus it was the part of the bolt between the flange and the hanger. The bolt was significantly tapered at the end to an approximately ¾-inch diameter. This was similar to that which is seen when a bolt is stretched in a tensile tester (Figure 6.11), which causes tensile strain. When this piece was examined using a stereomicroscope, the tapered edge showed a characteristic tensile cup-and-cone fracture as opposed to a fatigue failure. The surface of a tensile fracture fails along sheer planes which causes the surface to appear pitted or puckered when viewed with a stereomicroscope or SEM, as opposed to the linear fracture lines seen on the surface of a fatigue failure. These observations about the bolt led to a question regarding the source of the excess tension that caused this tapering and tensile fracture. Second, a steel plate or flange that acts as a retainer of the 1-inch diameter pin assembly in the hanger, was not flat and instead had a convex distortion. At this time, it was unknown to the scientist whether the steel plate was flat or convex when originally installed in the bridge. It was suspected that this plate was initially flat, and the convex deformation was caused by a considerable force putting pressure onto the 1-inch diameter pin. The source of this pressure was unknown at that time. To answer these questions, the scientist needed more information about the bridge design and construction, which prompted a trip to the bridge itself. The specific focus was to examine an adjacent section of the bridge that had not collapsed so that functioning bridge components could be observed in situ.

At the site of the bridge, the scientist was able to examine an undamaged pin and hanger assembly of an adjacent section of the bridge (Figure 6.12). The retainer plate was flat on the in-place bridge section. This indicated that there had been an internal force acting on the failed pin and hanger assembly that pushed out and deformed the steel plate retainer. To investigate the cause of the force that acted on the bolt and steel plate retainer, another section of the bridge was inspected which had a similar convex distortion of its steel plate retainer to that of the recovered failed parts that had been examined at the forensic lab. A hole was drilled through the deformed steel plate retainer on the bridge revealing excessive corrosion and debris. This confirmed that there was no nefarious cause for the bridge

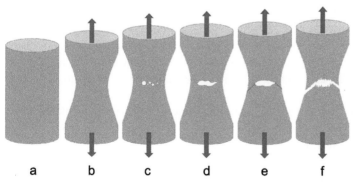

Figure 6.11 Formation of a cup-and-cone fracture: the initial cylinder (a), then as the two ends are pulled apart, the distortion of the material causes necking (b), the formation of microvoids (c), the merging of microvoids to form a crack (d), the propagation of the crack by shear deformation (e), and last the cup-and-cone fracture (f).

Figure 6.12 Dr. John Reffner examining the damage to the Myanus River Bridge. The hanger from the broken pin and hanger assembly can be seen behind Dr. Reffner, in the center of the photograph.

collapse (i.e., explosives), and instead the formation of rust and accumulation of debris over time caused the pin and hanger assembly to fail. When this component broke, this span of the bridge was free to collapse into the river as there were no fail-safe in its design to prevent this section from falling 70 feet into the Mianus River.

Despite having the answer to the cause of the bridge collapse within the first day after the event thanks to the microscopical analysis by a scientist with a prepared mind, an investigation continued for many months. This did not include participation from the forensic laboratory because it was clear that there were no criminal activities involved in the collapse. Ultimately, the final conclusion of the extensive investigation was the same as that provided by Dr. Reffner after less than 12 hours – the culprit was corrosion and excessive debris. The official Highway Accident Report stated on page 65:

> The National Transportation Safety Board determines that the probable cause of the collapse of the Mianus River bridge span was the undetected lateral displacement of the hangers of the pin and hanger suspension assembly in the southeast corner of the span by corrosion-induced forces due to deficiencies in the State of Connecticut's bridge safety inspection and bridge maintenance program.

There were two major consequences that resulted from the collapse of the Mianus River Bridge. First, were the immense monetary expenses to the State of Connecticut, which exceeded $16 million in direct costs.

> These direct costs include the work connected with the salvaging of the collapsed and damaged bridge structures; the purchase, installation and removal of a temporary span; costs of the retrofit of enlarged pins and hangers; the construction and installation of the replacement span.... It does not include the indirect costs such

as traffic control, detours, loss of tolls, litigation, etc. These indirect costs are still developing and cannot be determined precisely at this time.

(U.S. National Transportation Safety Board, 1984, p. 18)

Second, this collapse revealed a significant problem with bridge safety and maintenance programs, both in Connecticut and many other parts of the country.

The steel superstructure on the Mianus River Bridge was not kept clean. Pigeon excrement was piled up 6 to 10 inches on the bottom flanges of some of the steel. Not only did this add to the corrosive process (pigeon excrement contains urea, an ammonia salt), but it also discouraged the inspectors from walking the steel for closeup examinations of the pins and hangers. The ConnDOT maintenance policy did not call for the flushing of bridge superstructures. Steel should be kept free not only of bird excrement (particularly urea) but also of dirt, which can accumulate and hold moisture, which along with oxygen will cause corrosion of unprotected structural steel.

(Burnett, 1984, p. 3)

Once this was recognized, new measures were implemented which overhauled the safety and maintenance programs for bridges throughout the country.

References

Burnett, J. (1984, July 19). *Safety Recommendation(s) H-84-31 through -39.* National Transportation Safety Board.

U.S. National Transportation Safety Board. (1984, July 19). Highway Accident Report – Collapse of a Suspended Span of Interstate Route 95 Highway Bridge over the Mianus River Greenwich, Connecticut, June 28, 1983. *U.S. Department of Commerce.* National Technical Information Service.

6.4 Microcrystals Tests for Drugs Using the Chemical Microscope (PLM)

Gary J. Laughlin, Ph.D.

McCrone Research Institute, Chicago, IL, USA

A considerable problem facing forensic laboratories is the tremendous number of suspected illicit drug exhibits submitted for analysis, resulting in a backlog of samples requiring identification. This backlog prevents the timely resolution of cases within the criminal justice system. Chemical microscopy presents a unique solution to this problem.

The microscope has been available for hundreds of years, and its analytical results have been used in laboratory testing and courtroom cases for more than a century. What should be an obvious choice in any laboratory and a universal symbol of science, the light microscope is often regarded as passé by analytical chemists, who are indoctrinated by academia and the workplace with gas chromatography-mass spectrometry (GC-MS), X-ray diffraction (XRD), high pressure liquid chromatography (HPLC), Fourier transform-infrared (FT-IR) spectroscopy, and other "high tech" newer and usually non-microscopical techniques. The light microscope is an economical and versatile tool for solving a variety of problems by experienced scientists working in laboratories with a range of resources – from the most well-equipped to those on a restricted budget. The fact that the requisite sample size for microscopical analysis rarely exceeds milligram quantities is a bonus, with results often as good or better at the microgram level and an added advantage over other instrumentation. Yet the light microscope is usually disregarded or passed over in favor of other methods in most chemistry laboratories today.

The chemical microscope, also known as the polarized light microscope (PLM), is used for forensic science purposes, with the major part of the analysis in a subcategory of microchemistry called the microcrystal test for the screening of drugs and controlled substances. This test involves dissolving a very small portion of a suspected substance in a suitable solvent (usually aqueous) then adding a precipitating reagent under conditions that promote crystallization of microscopically unique material. When viewed directly through the microscope, a proficient microscopist (with a prepared mind) can positively identify, illustrate, and photographically document the results from tests for many controlled substances in a matter of minutes (Figures 6.13 and 6.14).

Specific microcrystal test methods, introduced by Professor Chamot at Cornell University, include: I, a drop of chemical reagent solution allowed to flow into a test drop with a specific drug or element, II a solid fragment of the chemical reagent added directly to a test drop, and III, a reagent drop manually drawn across a dried test drop solution (Chamot and Mason, 1940). Microchemical tests use proven technology and the PLM, a technically sound instrument, reliable, and accurate but generally regarded as subjective. Any doubt to its objectivity can be eliminated by additional modern research with the microscope, new discoveries of tests for newly developed substances, and better characterization and accessibility to each of the tests, reagents, their sensitivity, and specificity.

(A)

(B)

Figure 6.13 (A) A microcrystal test for the drug pseudoephedrine using a gold chloride reagent. Crystals form highly birefringent, thin, branching dendrites and combs; crossed polars. (B) Drawing of resultant pseudoephedrine microcrystals based on the research microscopist's direct observations (not to scale). Courtesy of McCrone Research Institute from the Microcrystal Compendium.

The real defense for microscope objectivity, reliability, and accuracy is to point out that the microcrystals obtained in a controlled test are unique to a given substance morphologically and optically through additional crystallographic properties, which can be observed, measured, and recorded directly in real time. Also of importance, microcrystal tests are applicable to the majority of controlled substances, including the most frequently encountered drugs, like cocaine and methamphetamine, and new designer drugs such as synthetic cathinones, also known as "bath salts" (McCrone Research Institute, 2021; Chamot and Mason, 1940; Clarke, 1969; Fulton, 1969; Horwitz, 1980; Julian and Plein, 1981; Ruybal, 1986; Silletti, 2015; Sparenga et al., 2020a, 2020b, 2020c, 2022a, 2022b; Stephenson, 1921). With all of these illicit drug compounds being organic, using the microscope and microcrystal tests, the molecule itself is identified rather than elements or functional groups. In addition, an

(A)

(B)

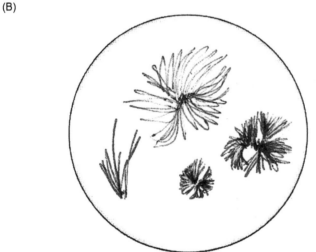

Figure 6.14 A microcrystal test for the drug alprazolam using a gold bromide with sulfuric acid and acetic acid reagent. Crystals form rosettes of needles and blades, burrs, fans, and sheaves; plane polarized light. (B) Drawing of resultant alprazolam microcrystals based on the research microscopist's direct observations (not to scale). Courtesy of McCrone Research Institute from the Microcrystal Compendium.

accomplished microscopist can recognize mixtures and identify many excipients and adulterants solely through the use of the PLM. They are trained in how to recognize and reduce the likelihood of interfering reactions, and they practice proven microscopical extraction techniques for illicit drugs and pharmaceuticals in various formulations and delivery mechanisms, including tablets, oral solutions, and transdermal patches.

The microcrystal test has experienced a resurgence not seen for decades, primarily through the efforts of Dr. Walter C. McCrone and McCrone Research Institute in Chicago, which he and his wife/fellow chemical microscopist Mrs. Lucy B. McCrone founded in 1960.

In 2015, **McCrone Research Institute** published *A Modern Compendium of Microcrystal Tests for Illicit Drugs and Diverted Pharmaceuticals* (the *Microcrystal Compendium*), containing 42 PLM microcrystal tests for 19 different drugs for which microcrystal tests using various reagents had been known but not each completely developed or compiled together or readily available in a useable modern format (McCrone Research Institute, 2021). The *Microcrystal Compendium* describes in detail the microcrystals formed from each test and includes test methods, reagent formulations, photomicrographs, morphology illustrations, optical properties, other drug interactions, and infrared microspectroscopy of the resultant microcrystals. It is available to view, search, and download for free on www.mccroneinstitute.org.

Since publication of the *Microcrystal Compendium*, these tests for illicit drugs have had a measurable and positive impact in the forensic science community and achieved the desired result for reliable analytical methods to assist crime laboratory scientists and other researchers in their work. As early as 2016, various federal, state, and municipal crime laboratories began incorporating, and in some instances reincorporating, microcrystals tests into their casework. McCrone in its courses and universities with chemistry or forensic science graduate studies programs have begun to use the *Microscopical Compendium* as a textbook or in laboratory exercises as part of their curriculum.

McCrone Research Institute offers their microscopy course, Microcrystal Tests for Illicit Drugs and Diverted Pharmaceuticals, to students worldwide employed as a microscopist or trace/physical evidence analyst in forensic crime laboratories. During the global COVID-19 pandemic, McCrone began presenting all of its illicit drug research through three online courses, including microcrystal tests for amphetamine, BZP, clonazepam, cocaine, codeine, diazepam, ephedrine, heroin, hydrocodone, hydromorphone, MDMA, methadone, methamphetamine, oxycodone, PCP, and pseudoephedrine. Also offered is a course in the extraction techniques suitable for a variety of pharmaceutical delivery mechanisms such as tablets, oral solutions, and transdermal patches. All of these courses, online and virtual, include basic and advanced microscopical techniques using the PLM with the microcrystal tests taken directly from the *Microcrystal Compendium* and additional research published in 2020–2022 **on** *New Microcrystal Tests for Controlled Drugs, Diverted Pharmaceuticals, and Bath Salts (Synthetic Cathinones)*, which contains microcrystal tests and reagents for several additional drugs for which there were previously no known microcrystal tests (Sparenga et al., 2020a, 2020b, 2020c, 2022a, 2022b).

The Minnesota Bureau of Criminal Apprehension (BCA) exemplifies the most notable example of how successful these tests have been in real-world scenarios. In January 2016, two months after the publication of the *Microcrystal Compendium*, the BCA implemented microcrystal tests at the Hennepin County laboratory, which serves Minneapolis. By September 2016, the BCA, after visiting other laboratories and having one of their analysts take the McCrone Research Institute microscopy course, were able to *completely* eliminate their backlog and reduce their turnaround time for cocaine, methamphetamine, and heroin from 6 months to 30 days. In the fall of 2017, they began work on phasing in selected microcrystal tests to additional counties and, as a result, 70% of the counties in Minnesota actively performed these tests with no false positives. In 2018, the BCA together with the Minnesota Department of Public Safety received the Better Government Award from their governor for great customer service and improving drug evidence analysis.

Becoming an accomplished microscopist takes time and training, but no more than what is required for proficiency in most other highly technical fields. Microscopy courses are offered now in-person and online, microscopes are readily available and relatively inexpensive when compared to other instrumentation, and the cost and speed of analysis is proven to save time and money. Any scientist who learns and practices these techniques will no doubt gain confidence in the specific, reproducible, and accurate results attained through microscopy.

References

ASTM E1968-19: Standard Practice for Microcrystal Testing in Forensic Analysis of Cocaine. (2019). West Conshohocken, PA: ASTM International. 2019.

ASTM E1969-19. Standard Guide for Microcrystal Testing in Forensic Analysis of Methamphetamine and Amphetamine. (2011). West Conshohocken, PA: ASTM International. 2019.

Chamot, E. M., & Mason, C. W. (1940). *Handbook of Chemical Microscopy* (Vol. *II*). Wiley. Reprinted by McCrone Research Institute, 1989.

Clarke, E. G. C. (1969). *Isolation and Identification of Drugs* (Vol. *1*). The Pharmaceutical Press.

Fulton, C. (1969). *Modern Microcrystal Tests for Drugs.* Wiley-Interscience.

Horwitz, W. (Ed.). (1980). *AOAC Official Methods* (13 ed.). Association of Analytical Chemists.

Julian, E. A., & Plein, E. M. (1981). Microcrystalline Identification of Drugs of Abuse: The 'White Cross Suite'. *Journal of Forensic Science, 26*(2), 358–367.

McCrone Research Institute (2021). A Modern Compendium of Microcrystal Tests for Illicit Drugs and Diverted Pharmaceuticals. Published online 2021. Accessed July 20, 2023. https://www.mccroneinstitute.org/uploads/A_Modern_Compendium_of_Microcrystal_Tests.pdf

Ruybal, R. (1986). Microcrystalline Test for MDMA. *Microgram, 19*(6), 79–80.

Silletti, D. (2015). Development of a Microcrystal Test for the Detection of Clonazepam. *The Microscope, 63*(2), 51–56.

Sparenga, S. B., Laughlin, G. J., King, M. B., & Golemis, D. (2020a). New Microcrystal Tests for Controlled Drugs, Diverted Pharmaceuticals, and Bath Salts (Synthetic Cathinones): Alprazolam and Butylone. *The Microscope, 68*(1), 17–32.

Sparenga, S. B., Laughlin, G. J., King, M. B., & Golemis, D. (2020b). New Microcrystal Tests for Controlled Drugs, Diverted Pharmaceuticals, and Bath Salts (Synthetic Cathinones): Mephedrone and Methylone. *The Microscope, 68*(2), 80–93.

Sparenga, S.B., Laughlin, G.J., King, M.B., & Golemis, D. (2020c). New Microcrystal Tests for Controlled Drugs, Diverted Pharmaceuticals, and Bath Salts (Synthetic Cathinones): MDPV and 4-MEC. *The Microscope, 68*(3–4), 156–171.

Sparenga, S.B., Laughlin, G.J., King, M.B., & Golemis, D. (2022a). New Microcrystal Tests for Controlled Drugs, Diverted Pharmaceuticals, and Bath Salts (Synthetic Cathinones): Alpha-PVP. *The Microscope, 69*(2), 84–91.

Sparenga, S.B., Laughlin, G.J., King, M.B., & Golemis, D. (2022b). New Microcrystal Tests for Controlled Drugs, Diverted Pharmaceuticals, and Bath Salts (Synthetic Cathinones): Tramidol and Zolpidem. *The Microscope, 69*(4), 175–187.

Stephenson, C. H. (1921). *Microchemical Tests for Alkaloids.* Charles Griffin & Co., Ltd..

7

Know Your Limitations

Introduction

Everything has its limits, including people and microscopes. Knowing one's limitations is incredibly difficult for some people, but is an essential component to efficiently solve problems. Similarly, a microscope is limited in resolution but not in magnification. Attempts at exceeding the resolution limits of a microscope result in empty magnification, or blurred images with no additional useful information beyond that contained in the initially magnified images. Attempts at problem solving using an individual beyond their capabilities may likewise lead to empty results, unsupported conclusions, and failure.

When solving problems, one must recognize one's own limitations and be able to ask for help when needed or when one has reached the boundaries of the individual or the tool being used. This chapter includes case studies which demonstrate the advantages of recognizing one's limitations and the creative problem solving that may occur by seeking help, thinking outside of the box, and exploring alternative pathways. As stated by Shirley Conran, a British novelist and journalist, "Self-confidence is the most important thing, and this comes from identifying your goals, knowing your limits and roping in all the help you can get."

Solving Problems with Microscopy: Real-life Examples in Forensic, Life and Chemical Sciences, First Edition.
Edited by John A. Reffner and Brooke W. Kammrath.
© 2024 John Wiley & Sons Ltd. Published 2024 by John Wiley & Sons Ltd.

7.1   Why Does Guercino's *Samson Captured by the Philistines* Have a Grainy Surface Texture in Some Paint Passages?

Sylvia. Centeno, Ph.D.

Department of Scientific Research, The Metropolitan Museum of Art, New York, NY, USA

Scientists, conservators, and art historians interrogate works of art by looking at them with the naked eye first, followed by a microscopic examination under different illuminations and using non-invasive analytical and imaging techniques such as X-radiography, infrared reflectography, macro X-ray fluorescence, and multispectral imaging. In some cases, important questions relevant to the artistic technique, technological process, conservation, and/or preservation of the work of art in question cannot be answered with these techniques alone, so professionals in museums and other cultural institutions take micrometer-sized samples that, in the case of paintings, generally include the full stratigraphy of the paint, from the ground preparation to the uppermost paint layer or varnish. An important question that conservation scientists have to ask themselves is whether the condition of the work permits taking samples from certain locations. Ideally, samples should be removed from areas where there is loss or a crack so as not to induce damage. However, it is not uncommon to find that a work of art is in such a pristine condition that sampling is not possible. Other issues relating to the condition of the artwork, such as past restoration campaigns and the possible contamination of the edges or other areas of the work due to handling, must be taken into account. Other questions are whether these samples are likely to provide the information needed and are representative of the phenomenon to be studied (Centeno, 2021).

Samson Captured by the Philistines, in the Collection of The Metropolitan Museum of Art in New York, was painted in 1619 by the Italian artist Guercino (Giovanni Francesco Barbieri, 1591–1666) (Figure 7.1). The painting, executed in a drying oil medium over a

Figure 7.1   *Samson Captured by the Philistines*, 1619. Oil on canvas. Guercino (Giovanni Francesco Barbieri; Italian, 1591–1666). 191.1 × 236.9 cm. The Metropolitan Museum of Art, Gift of Mr. and Mrs. Charles Wrightsman, 1984 (1984.459.2) (The Metropolitan Museum of Art/Wikimedia Commons/Public domain).

canvas support, depicts a dramatic scene from the Old Testament in which Samson, the strong man in Hebrew resistance against the Philistines, whose strength came from his hair, was shorn by his deceitful lover, Delilah, and then attacked by the Philistines, who blinded him (The Metropolitan Museum of Art, 2020).

This painting shows, when viewed under the magnification of a binocular stereomicroscope, lumps that give the paint surface a remarkable grainy texture (Figure 7.2). These lumps may be the result of the intentional addition by the artist of a coarsely ground material, a pigment or sand grains for example, to the paint mixture. However, grainy surface textures in oil paintings may also be due to a deterioration process called soap formation.

Soap formation in oil paintings is one of the most pervasive forms of deterioration and occurs when saturated fatty acids in the oil binding medium react with heavy metals in the pigments, driers, or other paint additives, such as lead and zinc, to produce heavy metal carboxylates, also called soaps. This process may cause the formation of disfiguring protrusions that grow and eventually may break through the paint surface, increased paint transparency, flaking, delamination, surface crusts, and the appearance of crater-like holes when the soap protrusions are abraded (Centeno & Mahon, 2009; Chen-Wiegart et al., 2017; Hale et al., 2011; Noble, 2019; Osmond, 2019). High temperature and ambient moisture, as well as aqueous solutions and organic solvents used in conservation treatments, are known to trigger and/or accelerate soap formation (Baij et al., 2018; Catalano et al., 2019; Noble, 2019; Osmond, 2019). Therefore, an in-depth study of the paint stratigraphy and composition is crucial, not only for understanding the artists' methods but also for making recommendations for the conservation and preservation of the works of art.

To determine the cause of the surface texture in *Samson Captured by the Philistines*, two microscopic samples were taken. The first one was removed from a grain just at the top of the paint surface in Samson's leg and was mounted as a cross-section in a synthetic resin and polished (Figure 7.3a). The second sample, taken from the floor area in the lower right of the composition, includes paint deeper below the surface than the first sample. This second sample was examined and analyzed without mounting it (Figures 7.4a and b).

Raman microspectroscopical analysis of the cross-section revealed that the lumps forming the remarkable grainy texture visible throughout the surface are in fact coarse

Figure 7.2 Detail of Guercino's *Samson Captured by the Philistines* photographed using a stereomicroscope at 3.5-times magnification with raking light (i.e., oblique illumination), showing the textured paint surface on Samson's right leg.

(a)

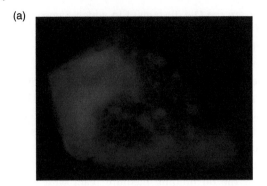

(b)

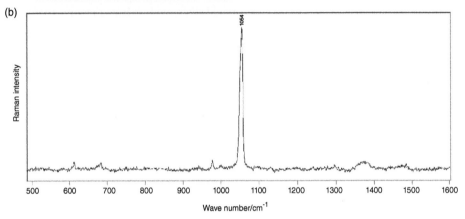

Figure 7.3 (a) Cross-section of a paint sample removed from a surface grain in Samson's leg as viewed with a reflected light microscope at 200-times magnification, showing a coarse particle, approximately 20–30 micrometers in diameter, identified as the pigment lead white by Raman spectroscopy, surrounded by more finely divided ochre and lead white particles. The ground layer is missing in this cross-section. (b) Raman spectrum acquired of the coarse white particle in the cross-section presented in (a), showing the strongest characteristic band for lead white at *ca.* 1054 cm^{-1} (λ_0 = 785 nm).

particles of the pigment lead white surrounded by more finely divided ochre pigment particles (Figure 7.3b). However, Raman analysis of the materials deep within the paint layer in the second sample, where areas of the pigment particles appear to be more transparent, revealed that some of the lead white has in fact undergone saponification (Figure 7.4c). The appearance of these areas when examined using a stereomicroscope and a reflected visible light microscope did suggest the presence of soaps however the Raman microspectral analysis was necessary to firmly identify the degradation. From the results obtained in these two samples, it is possible to state that even though saponification was not detected in the surface of the painting, the deterioration process is already taking place deep in the paint stratigraphy. The presence of lead soaps in just one sample taken from such a large painting attests to the widespread occurrence of the phenomenon in oil paintings.

As was the practice of many painters during the Baroque period, Guercino painted *Samson Captured by the Philistines* on a dark brown ground preparation. He built the

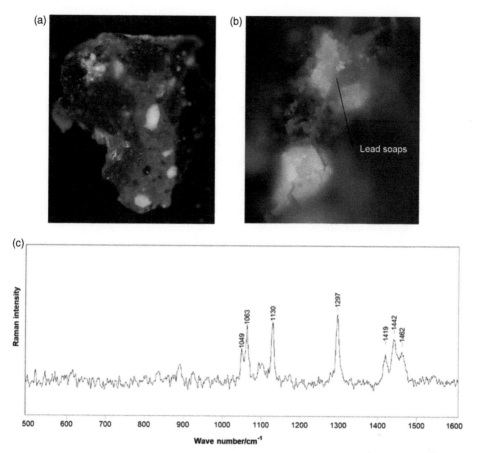

Figure 7.4 (a) Reflected light photomicrograph of the top of a microscopic sample removed from the bumpy surface at the lower right of the painting, showing coarse lead white pigment particles mixed with a more finely divided ochre pigment. Photomicrograph taken using reflected visible illumination, original magnification 100-times. Bottom view of the same sample presented in (a), showing white areas approximately 20 micrometers in diameter and slightly more transparent that the white particles in (a), where lead soaps that resulted from the reaction of the lead white pigment with fatty acids from the oil binder were identified by Raman spectroscopy.
(b) Photomicrograph taken using reflected visible illumination, original magnification 100-times.
(c) Raman spectrum acquired in a transparent white area in the sample shown in (b), with bands characteristic of lead soaps (Robinet & Corbeil, 2003; λ_0 = 785 nm) (adapted from [10]).

different paint passages with thin paint layers that he applied in scumbles, or light semi-opaque mixtures, over the dark ground to achieve the dramatic lighting so characteristic of the Baroque. The value relationships of many paintings of this period have shifted considerably, the dark ground becoming more dominant as the upper paint layers grow more transparent over time. The result is an increased contrast between the light and dark passages. It is possible that one of the causes of this growing transparency is the saponification throughout the paint layers of pigments containing heavy metals, as demonstrated using microscopical analysis in *Samson Captured by the Philistines* (Centeno & Mahon, 2009).

References

Baij, L., Hermans, J. J., Keune, K., & Iedema, P. D. (2018). Time-dependent ATR-FTIR spectroscopic studies on solvent diffusion and film swelling in oil paint model systems. *Macromolecules, 51*, 7134–7144. https://doi.org/10.1021/acs.macromol.8b00890.

Catalano, J., Murphy, A., Yao, Y., Zumbulyadis, N., Centeno, S. A., & Dybowski, C. (2019). Understanding the dynamics and structure of lead soaps in oil paintings using multinuclear NMR. In F. Casadio, K. Keune, P. Noble, A. van Loon, E. Hendriks, S. A. Centeno, & G. Osmond (Eds.), *Metal Soaps in Art-Conservation & Research* (pp. 69–84). Cultural Heritage Science Series. Springer.

Centeno, S. A. (2021). The spectroscopic study of pigments and binders in works of art. In J. M. Madariaga (Ed.), *Analytical Strategies for Cultural Heritage Materials and their Degradation*. The Royal Society of Chemistry.

Centeno, S. A., & Mahon, D. (2009). The chemistry of aging in oil paintings: metal soaps and visual changes. *The Metropolitan Museum of Art Bulletin, 67*(1), 12–19. http://www.jstor.org/stable/40588562

Chen-Wiegart, Yc. K., Catalano, J., Williams, G., Murphy, A., Yao, Y., Zumbulyadis, N., Centeno, S. A., Dybowski, C., & Thieme, J. (2017). Elemental and molecular segregation in oil paintings due to lead soap degradation. *Scientific Reports, 7*, 11656. https://doi.org/10.1038/s41598-017-11525-1.

Hale, C., Arslanoglu, J., & Centeno, S. A. (2011). *Studying Old Master Paintings. Technology and Practice. The National Gallery Technical Bulletin 30th Anniversary Conference Postprints* (Marika Spring, Ed.). Archetype Publications and The National Gallery.

The Metropolitan Museum of Art. https://www.metmuseum.org/art/collection/search/436603. Accessed August 4, 2020.

Noble, P. (2019). A brief history of metal soaps in paintings from a conservation perspective. In F. Casadio, K. Keune, P. Noble, A. van Loon, E. Hendriks, S. A. Centeno, & G. Osmond (Eds.), *Metal Soaps in Art-Conservation & Research* (pp. 1–22). Cultural Heritage Science Series. Springer.

Osmond, G. (2019). Zinc soaps: an overview of zinc oxide reactivity and consequences of soap formation in oil-based paintings. In F. Casadio, K. Keune, P. Noble, A. van Loon, E. Hendriks, S. A. Centeno, & G. Osmond (Eds.), *Metal Soaps in Art-Conservation & Research* (pp. 25–46). Cultural Heritage Science Series. Springer.

Robinet, L., & Corbeil, M. C. (2003). The characterization of metal soaps. *Studies in Conservation, 48*(1), 23–40. https://doi.org/10.1179/sic.2003.48.1.23

7.2 The Secrets of Hair

John A. Reffner, Ph.D.[1] and Brooke W. Kammrath, Ph.D.[2,3]

[1] John Jay College of Criminal Justice, New York, NY, USA
[2] University of New Haven, West Haven, CT, USA
[3] Henry C. Lee Institute of Forensic Science, West Haven, CT, USA

Hair is one of the most common traces recovered at the scenes of crimes due to its ubiquity, durability, and ability to be transferred without notice. Consequently, when recovered, hair is an important piece of forensic evidence for associating a suspect or victim with a crime scene or each other (Bisbing and Saferstein, 1982; Gaudette and Robertson, 1999; Petraco et al., 1988). According to the American Academy of Dermatologists, an individual will lose approximately 50–100 strands of hair per day. Although the majority of loss occurs during hair washing and grooming, hair is readily transferred through a range of activities, such as the commission of violent crimes. However, the utility of forensic hair examinations has been seriously questioned for more than a decade and continues to be a controversial topic (National Research Council, 2009; President's Council of Advisors on Science and Technology, 2016). At the core of these criticisms are issues regarding the evidentiary significance of hair evidence, and specifically questions regarding the limitations of microscopical comparisons and identification of features related to postmortem decomposition. The failure of forensic hair examiners to understand and effectively communicate the limitations of microscopical hair analysis is the pervasive problem.

Microscopical hair comparisons are complex due to the various microstructures within a hair. There are three principal regions of a hair: the root (or proximal end), the shaft, and the tip (or distal end). The shaft is composed of 3 distinct layers (like a pencil), with one layer wrapped around the other: the cuticle (a transparent scale structure covering the hair exterior), the cortex (the main body of the hair shaft), and the medulla (a column of air running through the center of the shaft, which may be filled with cells) (Figure 7.5). Each of these have diverse but characteristic morphologies with a total of 30–50 microscopic features used for the comparison of two hairs. A detailed examination using a comparison

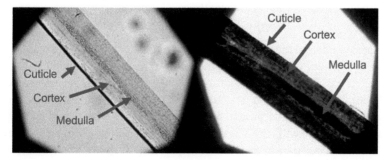

Figure 7.5 Brightfield photomicrographs of human hair as viewed with a 20-times objective, with the medulla, cuticle, and cortex labeled.

microscope that systematically describes all microscopical features greatly reduces the subjectivity of this analysis. As indicated in Chapter 3, images are data too.

In 2012, the FBI initiated a review of microscopical hair comparison analysis in response to three exonerations in the District of Columbia federal courts. The wrongful convictions of Santae Tribble, Donald Eugene Gates, and Kirk Odom were due, in part, to flawed testimony of forensic hair examiners. These errors included both misidentification of these individuals as being the source of evidence hairs which were shown to be excluded through mitochondrial DNA analysis and also exaggerated statements regarding the significance of an association through microscopical comparison. Of the 23,547 cases initially reviewed where testimony on microscopical hair comparisons was given, 3499 met the criteria for in depth analysis by the review board (ABS Group Report, 2019, p. 17). The five criteria for inclusion were (1) a "positive and probative" result from the microscopical hair examination, (2) conviction of the defendant, (3) the hair evidence did not undergo DNA analysis, (4) evidence was submitted and analyzed by the FBI laboratory scientists before December 31, 1999, and (5) a report was written and submitted to the appropriate law enforcement agency (ABS Group Report, 2019, p. 17). Of the transcripts and reports recovered, 93% (450 of 484) of the testimonies and 49.5% (856 of 1729) of the reports contained "errors." In total, it was concluded that 31 of the 35 FBI microscopical hair examiners erred in either their testimonies, reports or both, demonstrating the systemic problem. However, one must consider what was deemed an "error" in these reviews. The 2012 FBI Review of microscopical hair comparison analysis defined the following three types of errors (Federal Bureau of Investigation Laboratory, 2012):

Error Type 1: The examiner stated or implied that the evidentiary hair could be associated with a specific individual to the exclusion of all others. This type of testimony exceeds the limits of the science

Error Type 2: The examiner assigned to the positive association a statistical weight or probability or provided a likelihood that the questioned hair originated from a particular source or provided a likelihood that the questioned hair originated from a particular source, or an opinion as to the likelihood or rareness of the positive association that could lead the jury to believe that valid statistical weight can be assigned to a microscopical hair association. This type of testimony exceeds the limits of the science

Error Type 3: The examiner cites the number of cases or hair analyses worked in the laboratory and the number of samples from different individuals that could not be distinguished from one another as a predictive value to bolster the conclusion that a hair belongs to a specific individual. This type of testimony exceeds the limits of the science

The most common error fell into the first error type, and included statements that a hair was a "match to" or "consistent with having originated from" a specific individual. The term "match" was perceived as having a connotation of conclusiveness, and its usage was stopped after the OJ Simpson trial in 1996. The issue with the second statement arose in 2012, and was deemed to be an error because it did not explicitly recognize the possibility of other individuals having hair with the same microscopical features. Although these were approved conclusion statements in their time, the potential misleading information they possess has been identified and fixed in currently used language. In February of 2018, the United States Department of Justice developed guidance documents for forensic scientists with regard to establishing uniform language for testimony and reports (The United States Department of Justice, 2020 and 2021). One specific to the forensic hair discipline was updated in 2019 (The United States Department of Justice, 2019), and includes three approved conclusions for forensic hair examinations: inclusion, exclusion, and inconclusive.

The more serious errors involved the misuse of statistics by forensic scientists. In one of the most preposterous examples, an examiner provided testimony that the odds of evidence hairs coming from an individual other than the defendant was approximately 1 in a quadrillion. This probability was erroneously calculated by taking 5000 to the 4th power based on their being 4 evidence hairs with each hair having an unsupported error rate of 1 in 5000 (Lentini, 2015). To date, there is no established error rate for microscopical hair comparisons. This is a complex problem where one of the more difficult aspects to account for is the reliance on the examiner's skill for recognizing meaningful differences in the microscopic details of structures of the hair, which is based on their education, training, and experience. Although an examiner's skill is dependent on his or her ability to make observations, this is an essential component of the scientific method and does not invalidate the results.

Microscopic hair examinations have the unique ability to reveal distinct features of hairs that can be used to tell its history. Further, there is the possibly for this information to be used to aid in the forensic reconstruction of a crime. For example, darkening of the shaft combined with bubbles indicates that a hair has been in or near a fire (Figure 7.6). Microscopy is an important tool for identifying if a hair has been dyed or bleach (Figure 7.7). In the case of a bludgeoning death, hair that has been crushed may be recovered from the weapon (Figure 7.8). Remnants of insect activity can be observed using a microscope and appears as small bites to the hair shaft, and can show that a recovered hair was at a location for a time period beyond that of a specific event or criminal action (Figure 7.9). These are just a few of the more common examples, but highlight the importance of microscopic

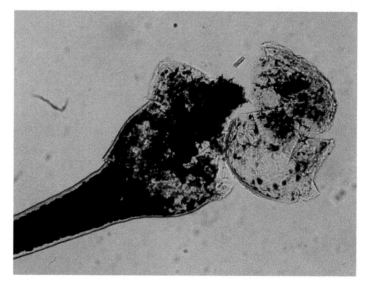

Figure 7.6 Brightfield photomicrographs of a human hair that has been singed, as viewed with a 10-times objective.

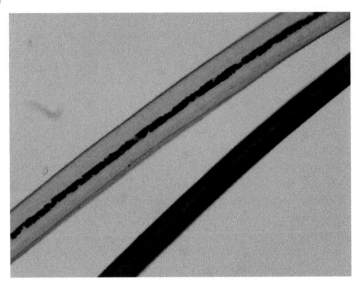

Figure 7.7 Brightfield photomicrographs of a single human hair that has been dyed, as viewed with a 10-times objective. This hair did not have a distinct dye line, and instead gradually faded from brown to light blond.

A   B

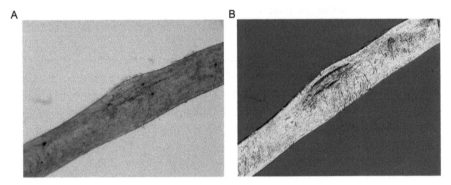

Figure 7.8 Brightfield (A) and PLM (B) photomicrographs of a human hair that has been crushed, each viewed with a 10-times objective.

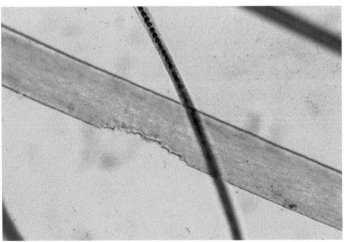

Figure 7.9 Brightfield photomicrograph of a human hair with insect damage, collected from the scene of a homicide investigation, as viewed with a 20-times objective. Also present in the image is a fine animal hair (fur), vertically oriented in the middle of the photomicrograph.

observations in hair examinations. There is no other analytical method capable of providing this valuable information about hairs.

Postmortem root banding (PMRB) is a distinctive hair feature that has received a large amount of attention due to its importance in high profile cases, such as the Woodchipper Murder in Connecticut, the Casey Anthony trial in Florida, and the wrongful conviction of John Kogut in New York (*People v. Kogut*, 806 N.Y.S.2d 373 (N.Y. 2005)). PMRB is commonly defined as the appearance of an opaque microscopic band near the root area of hairs from a decomposing body (Figures 7.10, 7.11, and 7.12) (Seta et al., 1984a, 1984b). Other terms have been used to describe this decomposition artifact, such as dead man's root, dead man's ring, or putrid root. Three stages of postmortem hair decomposition have been described by Koch et al. (2013). The first is the appearance of some observable banding, the second is the presence of a distinct band (approximately 0.25 mm in length) roughly 2 mm from the top of the root sheath and 0.5 mm from root tip, and the third is a brush-like end with darkening of the "bristles" occurring after the hair has broken off from the root at the location of the PMRB. PMRB was first introduced to the scientific community by two Japanese scientists, Sato and Seta, at an Oxford University meeting of forensic scientists in 1984 (Seta et al., 1984a, September 18–25), and shortly thereafter presented at the biannual meeting of the California Association of Criminalists (Seta et al., 1984b). Its value in a forensic investigation is that a hair with PMRB came from a dead body, and its presence at a location must therefore be explained (Tafaro, 2000).

One of the biggest problems regarding PMRB is that limitations regarding its formation and characterization have not yet been fully investigated or understood (Kadane, 2015). A number of research studies have been published addressing specific questions regarding the formation and characterization of PMRB (Collier, 2005; Collins, 1996; Damaso et al., 2018; Domzalski, 2004; Donfack and Castillo, 2018; Hietpas et al., 2016; Koch et al., 2013; Linch and Prahlow, 2001; Richard et al., 2019; Roberts et al., 2017; Tridico et al., 2014); however, two fundamental issues remain which limit its evidentiary significance. The first

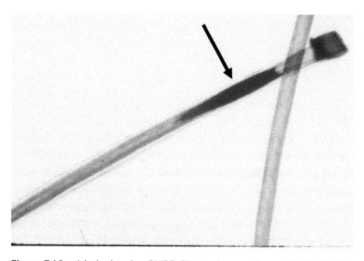

Figure 7.10   A hair showing PMRB. Photomicrograph provided by Dr. Peter R. De Forest.

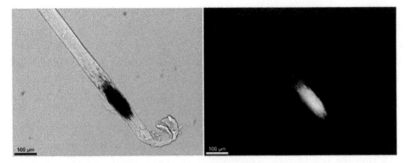

Figure 7.11 Transmitted light and darkfield (oblique) photomicrographs (left and right, respectively) of a known postmortem hair with PMRB. The dark coloring of the band in transmitted light while being bright due to light scattering in darkfield indicates that the band is filled with voids (which was confirmed with SEM). Elongated voids at the distal end of the band, seen the in the transmitted light photomicrograph, are characteristic features for PMRB identification (Hietpas et al., 2016/with permission from ELSEVIER).

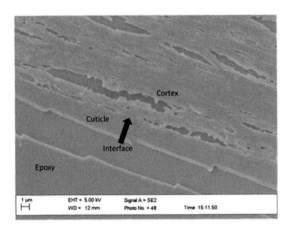

Figure 7.12 SEM photomicrograph of the PMRB in a known postmortem hair. The morphology of the voids in the cortex indicate that its structure is being ripped apart from the inside during the decomposition process. (Hietpas et al., 2016/with permission from ELSEVIER.)

unanswered question is concerned with the uniqueness of PMRB. Recent research has shown banding in antemortem hairs subjected to various environmental conditions (e.g., those exposed to aqueous, outdoor soil, or indoor shower environments) and chemical treatments (e.g., basic solutions) (Roberts et al., 2017). There are potentially morphological differences that have been proposed to differentiate these antemortem bands with PMRB (e.g., the location of the band with respect to the root tip, the presence of elongated voids near the edges of the distal end of the band, and the lack of damage or degradation of the cuticle), but more research is needed. The second relates to the timing for initial band formation. It is known that there is great variation between individuals with respect to postmortem band formation, both whether it forms at all and when a band will be first generated. Although it has been stated that band formation can begin 8 hours after death, to date, no reliable study has been conducted to test this conclusion. This is not the fault of the researchers, but instead it is due to limitations with experiments with humans. While experiments with donated bodies provide the best samples, information on the exact time of death is not necessarily available and time is required (usually a day or two) for the bodies to be made accessible to the researcher. These two issues demonstrate that we currently do not have enough information to conclusively define the scientific boundaries regarding the microscopical identification of PMRB that are necessary for determining it's the evidentiary significance. This does not imply that PMRB has no value in a forensic case, only that it is essential that written reports and testimony include the limitations of PMRB explicitly.

References

ABS Group Report. (2019). *Root and Cultural Cause Analysis of Report and Testimony Errors by FBI MHCA Examiners.* FBI Vault. https://vault.fbi.gov/root-cause-analysis-of-microscopic-hair-comparison-analysis/root-cause-analysis-of-microscopic-hair-comparison-analysis-part-01-of-01/view

Bisbing, R. E. (1982). The forensic identification and association of human hair. In R. Saferstein (Ed.), *Forensic Science Handbook* (pp. 184–221). Prentice Hall Regents.

Collier, J. H. (2005). Estimating the postmortem interval in forensic cases through the analysis of postmortem deterioration of human head hair (*Thesis*), Louisiana State University.

Collins, B. W. (1996). The effect of temperature and environment on post mortem morphology of human hair roots (*Thesis*). John Jay College of Criminal Justice, New York.

Damaso, N., Jones, K. F., Carlson, T. L., Fleming, J., Steadman, D. W., Jantz, L. M., ... Donfack, J. (2018). A study of the intrinsic variability and the effect of environmental conditions on the formation of a postmortem root band. *Forensic Science International, 293*, 63–69. https://doi.org/10.1016/j.forsciint.2018.09.032

Domzalski, A. C. (2004). The effects of environmental exposure on human scalp hair root morphology (*Thesis*). John Jay College of Criminal Justice, New York, NY.

Donfack, J., & Castillo, H. S. (2018). A review and conceptual model of factors correlated with postmortem root band formation. *Journal of Forensic Sciences, 63*(6), 1628–1633. https://doi.org/10.1111/1556-4029.13776

Federal Bureau of Investigation (2015). *FBI Testimony on Microscopic Hair Analysis Contained Errors in at Least 90 Percent of Cases in Ongoing Review.* Press Release. https://www.fbi.gov/news/pressrel/press-releases/fbi-testimony-on-microscopic-hair-analysis-contained-errors-in-at-least-90-percent-of-cases-in-ongoing-review

Federal Bureau of Investigation Laboratory. (2012, November 9). *Microscopic Hair Comparison analysis.* U.S. Department of Justice. https://www.mtacdl.org/attachments/CPE/Nelson/FBI_Limits_of_Science__%20Microscopic_Hair_Comparison.pdf.

Gaudette, B. D. (1999). Evidential value of hair examination. In J. Robertson (Ed.), *Forensic Examination of Hair*. Taylor and Francis. 1999.

Hietpas, J., Buscaglia, J., Richard, A. H., Shaw, S., Castillo, H. S., & Donfack, J. (2016). Microscopical characterization of known postmortem root bands using light and scanning electron microscopy. *Forensic Science International, 267*, 7–15. https://doi.org/10.1016/j.forsciint.2016.07.009

Kadane, J. B. (2015). Post-mortem root banding of hairs: A sceptical review. *Law, Probability and Risk, 14*(3), 213–228. https://doi.org/10.1093/lpr/mgv002

Koch, S. L., Michaud, A. L., & Mikell, C. E. (2013). Taphonomy of hair—a study of postmortem root banding. *Journal of Forensic Sciences, 58*, S52–S59. https://doi. org/10.1111/j.1556-4029.2012.02271.x

Lentini, J. (2015, September). It's Time to Lead, Academy News. *American Academy of Forensic Sciences, 45*(5), 5.

Linch, C. A., & Prahlow, J. A. (2001). Postmortem microscopic changes observed at the human head hair proximal end. *Journal of Forensic Science, 46*(1), 15–20. PMID: 11210902.

National Research Council. (2009). *Strengthening forensic science in the United States: A path forward*. National Academies Press.

People v. Kogut, 806 N.Y.S.2d 373 (2005).

Petraco, N., Fraas, C., Callery, F. X., & De Forest, P. R. (1988). The morphology and evidential significance of human hair roots. *Journal of Forensic Science, 33*(1), 68–76. https://doi. org/10.1520/JFS12437J

President's Council of Advisors on Science and Technology. (2016). Report to the President Forensic Science in Criminal Courts: Ensuring Scientific Validity of Feature-Comparison Methods. https://obamawhitehouse.archives.gov/sites/default/files/microsites/ostp/PCAST/pcast_forensic_science_report_final.pdf.

Richard, A. H., Hietpas, J., Buscaglia, J., & Monson, K. L. (2019). Timing and appearance of postmortem root banding in nonhuman mammals. *Journal of Forensic Sciences, 64*(1), 98–107. https://doi.org/10.1111/1556-4029.13810

Roberts, K. A., Garcia, L. R., & De Forest, P. R. (2017). Proximal end root morphology characteristics in antemortem anagen head hairs. *Journal of Forensic Sciences, 62*(2), 317–329. https://doi.org/10.1111/1556-4029.13286

Seta, S., Sato, H., Yoshino, M., & Miyasaka, S. (1984a, September 18-25). Morphological Changes of Hair Root with Time Lapsed After Death. In *Proceedings, 10th Triennial Meeting of the International Association of Forensic Sciences, Section on the Characterization of Human Hair*. IAFS, Oxford, UK.

Seta, S., Sato, H., Yoshino, M., & Miyasaka, S. (1984b) Morphological changes of hair root with the time lapsed after death (Proceedings of the California Association of Criminalists, Abstract #174, CAC). *Journal of Forensic Science Society, 4*, 338.

Tafaro, J. T. (2000). The use of microscopic postmortem changes in anagen hair roots to associate questioned hairs with known hairs and reconstruct events in two murder cases. *Journal of Forensic Science, 45*(2), 495–499.

Tridico, S. R., Koch, S., Michaud, A., Thomson, G., Kirkbride, K. P., & Bunce, M. (2014). Interpreting biological degradative processes acting on mammalian hair in the living and the dead: Which ones are taphonomic? *Proceedings of the Royal Society B: Biological Sciences, 281*(1796), 20141755. https://doi.org/10.1098/rspb.2014.1755.

The United States Department of Justice (2019). *Uniform language for testimony and reports for the forensic hair discipline*. U.S. Department of Justice. Retrieved from https://www.justice. gov/olp/page/file/1083686/download

The United States Department of Justice (2020). *Uniform Language for Testimony and Reports*. U.S. Department of Justice. Retrieved from https://www.justice.gov/olp/uniform-language-testimony-and-reports

The United States Department of Justice (2021). Uniform language for testimony and reports. Retrieved from https://www.justice.gov/olp/uniform-language-testimony-and-reports

7.3 A Connecticut Murder Case

John A. Reffner, Ph.D.[1] and Brooke W. Kammrath, Ph.D.[2,3]

[1] John Jay College of Criminal Justice, New York, NY, USA
[2] University of New Haven, West Haven, CT, USA
[3] Henry C. Lee Institute of Forensic Science, West Haven, CT, USA

On a summer night, a young woman left her home to walk to her job working the overnight shift at a dog kennel. She did not arrive to work that night, and was reported missing. A publicized search for the woman ended three days later when her body was discovered in a lake by a man and two boys going fishing. She had been stabbed to death.

An anonymous call was made to the state police describing a suspicious person burning something under a bridge. Dr. Henry Lee went to the crime scene, and collected the debris from the fire. Initial examination of the debris revealed a charred piece of orange fabric (Figure 7.13). The night of her disappearance, the young woman had been carrying her purse and a duffel bag containing a makeup case, a pair of jeans, and the kennel's employee t-shirt which was orange in color. The questioned charred fabric (166) and an exemplar shirt (55B) from the kennel's bulk supply was sent to the FBI for comparison.

FBI forensic analysis of the two fabrics consisted of identification of the weave structure of the textile and its fiber components. Although it was determined that the weave pattern of the two fabrics was consistent, the FBI forensic examiner made an inconclusive determination for two reasons. First, although the colors appeared similar, charring of the questioned fabric prevented a conclusive visual comparison. Second, and more importantly, the known shirt was a blend of cotton and polyester fibers, but only the cotton fibers were observed in the questioned sample. These results were submitted to Dr. Lee, but he was unsatisfied with the inconclusive conclusion. Consequently, the evidence items were sent to a consulting microscopist, Dr. John A. Reffner, for follow-up analysis.

The microscopist examined and documented the questioned and exemplar fabric swatches macroscopically and with a stereomicroscope (Figure 7.13). The stereomicroscopic

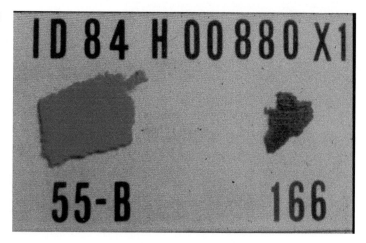

Figure 7.13 Photograph of the exemplar fabric from an orange Kennel employee t-shirt (55B, left) and the questioned charred fabric (166, right).

examination confirmed that the fabric weaves of the known and questioned samples had similar patterns, as reported by the FBI examiner. The questioned charred fabric was stiff, unlike the exemplar sample which had the expected flexibility of a t-shirt. Fortuitously, the microscopist had expertise in material science, with advanced knowledge in synthetic fibers and their manufacturing. The inflexibility of the charred fabric suggested to him that the polyester fibers may have melted. Polyester is a melt-spun fiber which is produced through the carefully controlled melting of polyester pellets that are then extruded through a spinneret to produce fibers of a specific cross-section and then stretched to orient the molecular structure to create stronger fibers. Although individual polyester fibers are flexible, a mass of polyester is not.

A polarized light microscopical examination of the two fiber samples revealed further information (photomicrographs of each are shown in Figure 7.14). Orange cotton and polyester fibers were easily seen in the samples of the exemplar. Cotton fibers are readily identified by their twisted ribbon-like morphology and undulating extinction when examined with cross-polarized light. Polyester fibers, like most synthetic fibers, have a uniform appearance, and can be identified by measuring their optical properties (e.g., parallel and perpendicular refractive index, birefringence, etc.). Only orange cotton fibers were observed in the questioned charred sample; however, a colorless second phase of material was seen dispersed between the cotton fibers. At this point, it was hypothesized that the colorless second phase was the result of the melting of the polyester fibers due to heat from the fire. To test this hypothesis, Fourier transform infrared (FT-IR) microscopy was used to analyze the unknown second phase. It should be noted that the FBI laboratory at this time did not have an FT-IR microspectrometer,

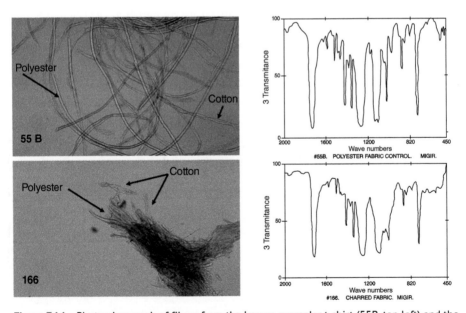

Figure 7.14 Photomicrograph of fibers from the known exemplar t-shirt (55B, top left) and the questioned charred fabric (166, bottom left), and the corresponding FT-IR spectra of the exemplar polyester fibers (top right) and the unknown second phase (bottom right).

so they were not able to do this analysis. With the FT-IR microspectrometer, the second phase seen in the questioned sample was identified as polyester (Figure 7.14). Finding polyester present in the questioned sample established that the two samples could have come from the same batch of t-shirts.

The microscopist went a step further. Upon examination of the samples with plane-polarized light, both the questioned and exemplar cotton fibers were observed to be dichroic. Dichroism is the phenomenon of an object displaying two different colors depending upon its orientation (parallel and perpendicular) when viewed in polarized light. A polarized visible light microspectrometer was used to analyze the questioned and known samples, and both had comparable dichroic spectra in both orientations of the linearly polarized light (Figure 7.15). This provided additional evidence of similarity between the two materials.

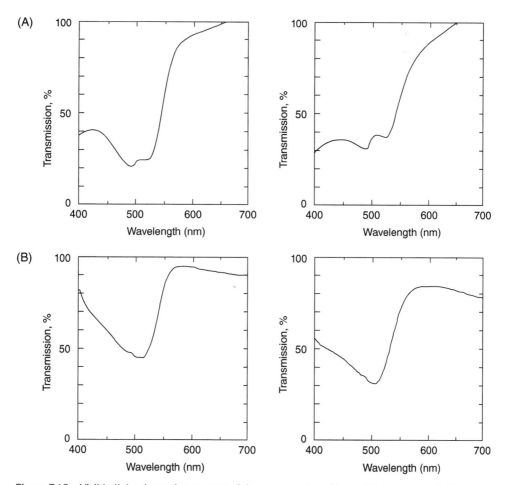

Figure 7.15 Visible light absorption spectra of the orange cotton fibers of the exemplar (left) and questioned (right) fibers, with the polarized light orientated parallel (A) and perpendicular (B) to the fiber's length.

The physical and chemical similarities of the charred fabric with the known exemplar shirt provided important evidence which was used to associate the victim with the fire. Other physical evidence, not discussed here, was used to associate the suspect with the fire, thus completing the evidence triangle (victim—suspect—crime scene). Although questions regarding the veracity of some of the other physical evidence have been raised in post-conviction efforts, the results of the analysis of the orange fabric remain unchallenged. Ultimately, the defendant was convicted of murder, sentenced to 60 years to life in prison, and released after serving 26 years due to a "sentence modification" made possible through the help of the Connecticut Innocence Project.

8

The Resonance Theory of Experiments

Introduction

What is meant by the title of this chapter, "The resonance theory of experiments"? By resonance we mean an enhancement, intensification, or enrichment of a concept or hypothesis through supplementary experimentation and information. This is consistent with the iterative process of the scientific method (see this book's Introduction), whereby findings made from initial observations are enhanced through subsequent testing and additional data. It is through this resonance theory of experiments that certainty in the validity of conclusions can be assured.

The concept of resonance is present in several disciplines, most notably music and the sciences (e.g., the molecular structure of matter). One of the most useful examples of the power of resonance can be seen in a molecule of benzene. Benzene is a common chemical found in petrochemicals and crude oil. Although it was used in pharmaceuticals and perfumes since the 16th century (as gum benzoin), it was first isolated and identified in 1825 by Michael Faraday. It was given the name benzin in 1834 by Eilhardt Mitscherlich, a German chemist, who distilled it from gum benzoin. Mitscherlich analyzed this substance, and determined that it was composed of six carbon and six hydrogen atoms (C_6H_6). However, based on current knowledge of bonding, it was expected that a molecule with 6 carbons should have 14 hydrogen atoms (C_6H_{14}). It wasn't until 1865 that French chemist, August Kekulé, had a dream about a snake biting its tail that the true structure of benzene was hypothesized: it is a six membered ring with alternating single and double bonds. Later it was recognized that the electrons were actually delocalized over three or more atoms, which increased the stability of the molecule. This power of resonance to increase the strength and stability of molecular bonds has had far reaching implications in the understanding of matter, without which the fields of pure and applied chemistry would not have been able to develop.

The case examples presented in this chapter demonstrate the value of the resonance theory of experiments, and how the microscope can be used to provide the resonating information which ensures the correct solution to a problem.

Solving Problems with Microscopy: Real-life Examples in Forensic, Life and Chemical Sciences, First Edition.
Edited by John A. Reffner and Brooke W. Kammrath.
© 2024 John Wiley & Sons Ltd. Published 2024 by John Wiley & Sons Ltd.

8.1 The Red Hooded Sweatshirt

John A. Reffner, Ph.D.[1] and Brooke W. Kammrath, Ph.D.[2,3]

[1] John Jay College of Criminal Justice, New York, NY, USA
[2] University of New Haven, West Haven, CT, USA
[3] Henry C. Lee Institute of Forensic Science, West Haven, CT, USA

A 13-year old girl did not return home from school one day in 1976. After the police were contacted at dusk, a large community search of the area commenced. The search ended around two hours later when the bloody body of the young girl was found in the bottom of an old foundation which was all that remained of an abandoned building. This was a popular location for neighborhood children to hang out because it was relatively remote and easy to access. Subsequent autopsy results revealed that she had died at around 4:00 p.m. with the cause of death being multiple skull fractures. It was thought that she was bludgeoned to death by a large rock weighing 26 pounds.

The crime scene was carefully preserved with access restricted to experienced police and scientific investigators who searched for physical evidence. The finding of a small tuft of white fibers with a red thread (Figure 8.1) was of particular significance. It was observed on the ground in the area near the body and was not covered with debris, thus making it out of place at this outdoor crime scene. Thus, it was collected as potentially vital evidence.

Police officers canvased the local area, and interviewed local residents and kids in the hopes of finding a witness to the murder. One 15-year old boy, an acquaintance of the victim, was interviewed in the presence of his parents. With knowledge of the questioned red thread that was recovered from the scene, the police asked if the boy had any red clothing. One of the concerns was whether the unknown fibers had evidentiary significance, thus police were collecting all red textiles from people in the area and from those who had searched for the body (e.g., volunteer firefighters). This young man volunteered his red hooded sweatshirt (Figure 8.2) to the police, who took it into evidence.

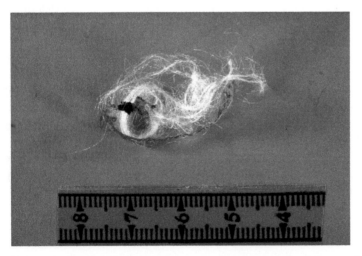

Figure 8.1 Documented stereomicroscope photomicrograph of the questioned white tuft of fibers with the red thread.

Figure 8.2 Photograph of the suspect's red hooded sweatshirt. The black circles drawn on the sweatshirt represent areas where blood traces were observed, tested, and documented.

The police submitted the red hooded sweatshirt to the state forensic laboratory, where it was compared it to the questioned red thread. The forensic examiner performed a visual comparison of the questioned red thread and the red hooded sweatshirt, and concluded that they were the same color. Further examination of the sweatshirt revealed the presence of a small amount of blood traces. At that time, DNA analysis was not available; however, ABO typing and other serological testing indicated that the blood was a "match" to the victim. As a result, this young boy became the primary suspect in the young girl's murder.

The value of multiple associations is detailed in Chapter 9, and was the impetus for sending the evidence fibers found at the scene and the known sweatshirt to a consulting microscopist for comparative analysis. Although the questioned red thread was visually similar in color to that of the known sweatshirt, this needed to be tested with microscopy and visible microspectroscopy. A fiber from the body of the sweatshirt was compared to those composing the questioned thread. There was no spectral match between these fibers, which prompted the microscopist to examine the entirety of the garment. A stereomicroscopical examination of the hooded sweatshirt revealed that it was composed of different red fibers in different areas of the garments. In addition to main garment fibers, different fibers were used in the manufacturing of various parts of the sweatshirt. For example, there were the knitted cuffs of the sleeves and waistband, the zipper tape, a thread used to attach the zipper to the garment, and the draw cord of the hood. None of these components of the sweatshirt had visible spectra consistent with that from the questioned thread at the scene.

Upon further reflection, the microscopist recognized that the red thread was found with a white tuft of acrylic fibers. Thus he focused on examining the red sweatshirt for a possible source of those white fibers. The only white component of the sweatshirt was a white label stitched to the back of the neck at the base of the hood (Figure 8.3). This was examined with a stereomicroscope, and it was noted that the label was frayed at one end. Polarized light microscopical analysis of the acrylic fibers composing the tag revealed that they had the same microscopical properties as the evidence white fibers. Additionally, he found another type of red thread that was used to secure the label to the sweatshirt. These fibers were

Figure 8.3 Stereomicroscope photomicrograph of the torn label of the suspect's red hooded sweatshirt, showing the frayed fibers from the tag.

compared to those of the evidence red thread, and were determined to be composed of the same fiber and dye via polarized light microscopy and visible microspectroscopy. This indicated that the two evidence fibers (the red thread and white tuft of acrylic fibers) could have originated from the same source as those used in the construction of the label and its stitching, or another source with the same chemical and physical properties of both fibers. It must be recognized that fibers are mass manufactured items, which may limit their evidentiary significance. However, when the microscopical and spectral matching are combined with the blood evidence and the frayed condition of the label, the microscopist's association of the evidence fibers to the suspect's sweatshirt was meaningful for building the prosecutor's case.

Unfortunately, it is not always about the physical evidence. During the trial, the defense was able to get the sweatshirt excluded by arguing seizure of the sweatshirt and other items violated the defendant's constitutional rights (Frisman, 1978). The defendant had not been informed of his right of refusal, claimed he was "intimidated by the badge," and the officer did not have a warrant to seize the sweatshirt or other items. It should be noted that at the time of the initial canvasing of the neighborhood and interaction with the defendant, the officer claimed that he was only interviewing potential witness and was not aware that the young man was a minor. By all accounts, the young man looked much older than his age as he was over 6 feet in height and had facial hair. As a consequence of the judge's decision, the prosecutor had to obtain a new warrant for the forensic testing of the sweatshirt. All of the blood evidence had been consumed in initial testing, and thus was not available for reanalysis. The microscopist was able to replicate his analysis and presented it at trial. However, it was not enough to secure a conviction in the absence of the other associative evidence, and the jury issued a verdict of not guilty. Was justice served?

Reference

Frisman, P. (1978 February 3). Murder suspect's rights were violated, judge rules. *Hartford Courant*. 1.

8.2 Florida Arson Case

John A. Reffner, Ph.D.[1] and Brooke W. Kammrath, Ph.D.[2,3]

[1] John Jay College of Criminal Justice, New York, NY, USA
[2] University of New Haven, West Haven, CT, USA
[3] Henry C. Lee Institute of Forensic Science, West Haven, CT, USA

On October 31, 1961, an explosion and fire destroyed The Three Winners Boys Shop, Inc. in Miami Beach, Florida. This store was owned by Elias and Ruth Rausch. When the fire department arrived at the scene, the building was engulfed in flames and an explosion had blown out all of the windows. Nothing in the initial investigation provided an explanation for the explosion, which suggested that the fire could have been intentionally set. Further, upon arrival at the scene, a fire truck had knocked down a garbage bin, and a red gallon container marked "gasoline" fell out. This container was seized by the fire department and sent to the Dade County Crime Laboratory. Additionally, an investigator, Terrence Howell of John A. Kennedy and Associates, collected six pieces of plate glass from the fire scene which contained an "oily residue" on the surface. These were sent to Polytechnic, Inc., an associated company of Walter C. McCrone that was owned by Marvin A. Salzenstein, a professional engineer in Chicago, Illinois. The purpose of the investigation was to determine if chemical analysis would reveal the presence of accelerants which may have been used to cause or aid the fire.

The six pieces of plate glass evidence were described by scientist John A. Reffner, in his report as:

> Exhibit A: Six, irregularly shaped pieces of plate glass ¼" thick and spottily covered with a black, brittle stain on the surfaces. A localized browning stain was also noticed on the two surfaces of the various pieces of glass, as well as on the broken edges.

After a macroscopic examination, stereomicroscopy was used to investigate the exhibit. The black deposit was brittle in nature and could be removed from the glass by chipping. A small sample was removed, mounted in an immersion oil, and examined by polarized light microscopy. Among the black soot particles, small birefringent particles were observed. These particles were clearly crystalline in nature, which informed the next step in examination: X-ray diffraction.

X-ray powder diffraction is an analytical technique which uses X-rays and their reflection from crystalline materials. The reflection of X-rays from crystals varies as to angle and intensity, depending on the structure or makeup of the crystals. Examination of the patterns of X-ray diffraction (XRD) measurements are used to identify crystalline material in both pure substances and mixtures. Analysis of the XRD pattern of the black coating revealed that it was a complex mixture of solids composed of two crystalline components in addition to the black soot. The identity of these two components was determined through comparison to the National Bureau of Standards (the predecessor for the National Institute of Standards and Technology, NIST) XRD database. The two components were identified as potassium chlorate and manganese dioxide. Potassium chlorate is a powerful oxidizing agent which will actively burn, or even cause to explode, organic materials such as sugar, wood, or charcoal. Manganese dioxide is a catalyst to the reaction; causing rapid and complete combustion. Both of these chemicals would have been available from a chemical supply house or hobby shop, the latter of which carry these materials as toy rocket fuels. This combination of chemicals is also used in the manufacturing of flares or fusees, which is a pyrotechnic that produces a bright light without an explosion that is commonly used for signaling distress or warnings.

A possible method an arsonist may use for the delayed ignition of a purposeful fire is to fill a natural rubber prophylactic contraceptive with sulfuric acid, and place it over a mound of the chemicals previously described (potassium chlorate and manganese dioxide). As the rubber is attacked by the sulfuric acid, a break soon occurs and causes the acid to flow in contact with the chemicals. A vigorous fire with extremely high temperatures immediately ensues, causing ignition of other combustibles or flammables present.

The black soot residue was also examined for the presence of petroleum hydrocarbons by Fourier transform infrared (FT-IR) absorption spectroscopy. FT-IR spectroscopy is an analytical method which measures the molecular vibrations of a chemical for its identification. The absorption pattern of infrared radiation by a compound can be unique for that particular compound. In this case, the residue on the glass was extracted with carbon tetrachloride. Petroleum hydrocarbons are soluble in carbon tetrachloride and this solvent does not absorb infrared radiation in the regions of interest. The infrared absorption spectrum of the extract was identified by bands at 2925, 2860, 1450, and 1360 cm^{-1} corresponding to a simple hydrocarbon product, resembling a petroleum distillate.

If this was a petroleum product such as gasoline, kerosene, or diesel, it would be composed of a mixture of simple hydrocarbon solvents. To test this hypothesis, gas chromatography equipped with a flame ionization detector (GC-FID) was employed to determine the exact composition of the simple hydrocarbons present in the black soot residue. Gas chromatography is a separation method where a mixture is introduced into a narrow column, and separation of the components is achieved based on their boiling points and chemical affinity. Further, the time it takes for the chemicals to pass through the column and reach the detector (at a given flow rate) is used to identify each of the components. In the early 1960s, this instrumentation was quite novel and special arrangements were made to provide access to a laboratory with this analytical instrument. This included the scientists making their own packed columns in copper tubes, which was not a trivial task.

Additionally, the scientists compared the results of the gas chromatography analysis of this residue to that of the liquid found in the red gallon container labelled "gasoline" at the fire scene. The sample from the red gallon container consisted of a pale yellow, mobile, oily liquid observed to have a characteristic odor of a petroleum distillate. It had a density of 0.807 grams per milliliter and a boiling point range from 140° to 300°C. Based on the results of the gas chromatography, which had multiple peaks corresponding to a range of hydrocarbons, both the crime scene liquid and the hydrocarbon in the black soot residue were identified as belonging to the kerosene-class of ignitable liquids. The liquid recovered from the crime scene was of the same chemical nature as the hydrocarbons found in the residue on the glass pieces, indicating that they could have a common source. Additionally, organic materials which may normally be expected in a men's shop (such as clothing, flooring, wood paneling, etc.) could not account for the hydrocarbons observed.

As a footnote, there were two cases associated with this event. The first was a civil trial, where three insurance companies were suing the defendant, Elias Rausch, in order to avoid paying for damages caused by the fire. Mr. Rausch wanted this case to be tried first so that there was no associated arson conviction on his record to influence the civil case decision. The defendant was determined by the civil jury to be responsible for "burning to defraud the insurer," and thus the plaintiffs were relieved of responsibility for delivering their insurance payout. Next, Mr. Rausch was tried in criminal court, and was found guilty of arson. This carried a penalty which included jail time.

In Florida, this case was the first in over a decade where a person was found guilty of arson. The scientific findings, which included both the ignition mechanism and presence of an accelerant, were critical evidence for achieving this conviction. Further, for decades after this event, all suspected arson cases in Dade County included the crime scene collection and submission to the forensic laboratory of pieces of sooted glass since it had proved to be key to solving the arson problem.

8.3 The Multimillion-Dollar Waterproof Failure

Andrew Anthony Havics, CIH, PE

pH2, LLC, Avon, IN, USA

For complex problems, multiple techniques, including multiple microscopical techniques, may be necessary to provide a definitive answer. An example of the integrative aspects of multiple methods (microscopical and non-microscopical) was a waterproofing failure case. A modified bituminous or modbit waterproofing material had been observed oozing out of the balconies of a facility. The facility consisted of six-story and seven-story condominium buildings. The balconies (where the oozing had occurred) consisted of cantilevered structural steel beams supporting a metal deck with concrete sloped to drains. Waterproofing was installed on the balconies and other concrete slabs at the facility. Above the waterproofing layer installation, set in a cementitious material was a trace heating coil system used to melt snow on the top side of the upper concrete pad. After installation the waterproofing started oozing/bleeding out at spots, including into the balcony rain and melt water drains.

One of the important aspects of any successful investigation is Louis Pasteur's saying *"chance favors the prepared mind"* (Chapter 6). For a failure case, the analyst should know and understand the following product information: specifications and properties, design criteria, manufacturing process, installation process and environment, and conditions of use. This information can be used to develop an informed plan of action.

8.3.1 Material Background

In general, the matrix, excluding the reinforcement, of a waterproof system of this type would be expected to consist of (Funkhouser, 1999):

Bituminous Materials: 65–85%, Asphalts, Tackifiers, Oils
Reinforcing Fillers: 15–5%, Calcium Carbonate, Magnesium Silicate, Mica
Thermoplastic Polymers: 20–10%, Styrene-Butadiene-Styrene (SBS), Styrene-Isoprene-Styrene (SIS), Styrene-Ethylene-Butadiene (SEBS), Styrene-Butadiene (SBR), Polyvinyl Chloride (PVC), Neoprene

The specific manufacturer's specifications indicated that, "Application of the membrane shall not commence nor proceed during inclement weather. All surfaces to receive the membrane shall be free of water, dew, frost, snow or ice…Application of membrane shall not commence nor proceed when the ambient temperature is below 0° F…Over its service life, do not expose membrane or accessories to a constant temperature in excess of 180° F (i.e., hot pipes and vents or direct steam venting, etc.)." The manufacturer also provided warnings concerning the installation: do not apply below 0° F or to damp, frosty or contaminated surfaces; do not exceed maximum safe operating temperature of 400° F. They also explained that to properly melt a rubberized asphalt, the material should be melted slowly in a double-jacketed kettle, should not exceed a temperature of 400° F, and should be continually agitated to maintain a consistent blend of the asphalt and SBS rubber. They described that if the waterproofing is improperly melted, it will produce areas of inconsistent product with sections having only high penetration asphalt without the stabilizing SBS rubber. This indicated that the thermoplastic polymer portion in this product was an SBS. In order to apply the waterproofing and allow it to level properly, the material was to be heated in a kettle. Asphalt—polymer mixes are known to have separation tendencies [as well as dispersion issues (Cash, 2004)] upon extended or high heating, hence the development of ASTM test method D7173 (ASTM, 2013). Based on field records, the material was applied during snowy weather at 19–35° F, and below 0° F. These were two environmental conditions that could have affected the material installation process and performance. Field installation notes, interviews, and testimony did not reveal any particular spatial pattern differences that would allow insight into the cause.

8.3.2 Initial Hypotheses and Sample Treatment

A couple of strong suspect causes emerged early in the field investigation: 1) overheating in a single kettle system and 2) overheating from concrete heating coil system installed later and adjacent (vertically) to the waterproofing. In order to begin evaluation, five large (approximately 4–20 inches each side) samples of the material were extracted and delivered to the lab. See Figures 8.4 and 8.5 for examples of samples collected from the field. Visual observations showed obvious differences within the same purportedly homogenous material. In particular was the solid versus semisolid character of the waterproofing as installed. There appeared to be a separation of two components differentiable by their liquidity (see Figure 8.5).

Based on this early data from the field and from initial observations, the initial hypotheses were considered and then potential tests and analyses were conducted to assist in determining the cause of the material issue. Part of the preparation of samples required taking new material (unused) and subjecting it to heating for various times and temperatures to

Figure 8.4 Sample 1 of waterproofing material consisting of a thick, solid mat in good condition.

Figure 8.5 Sample 3 of waterproofing material consisting of a separated fluid-like layer on top of a solid mat.

mimic potential field conditions for the installed waterproofing. Samples of new water-proofing material were heated from 200 to 500° F and the times varied from 30 to 1440 minutes (1/2 hour–3 days). Testing or analysis on exemplars is commonly required to verify the manufacturer and manufacture of the field material, verify the material properties prior to alteration or failure, and verify or refute the effects of suspect conditions that led to the alteration or failure.

8.3.3 Initial Analysis

The initial analyses included determination of the softening point by ASTM D36/D36M-09 (ASTM, 2013) of the field samples and a representative new sample out of the box. The results demonstrated that four of the five field samples had softening points well below the manufacturer specs of 190–230° F, whereas the new product was in range while the addition of higher heating and longer heating to the new product continued to reduce the softening point. The lower the softening point, the less viscose the material and thus the more likely to flow and "ooze." Short term heating in the lab, however, did not create full separation of product. These mixed softening point results did suggest that there was at least some extended heating resulting in separation of phases (Lu & Isacsson, 1997; Lu et al., 1999).

Initial testing also included analysis by gel permeation chromatography (GPC). GPC has been used to characterize bitumen and Styrene-Butadiene-Styrene (SBS) polymers, and to evaluate the aging and thermo-oxidative degradation of polymer-modified bitumen (PMB; Canto et al., 2006; Daranga, 2005; Lee et al., 2008; Linde & Johansson, 1992; Lu & Isacsson, 1997; Masson et al., 2003, 2005, 2006, 2008; Molenaar et al., 2010; Read & Whiteoak, 2003; Snyder, 1969; Wang et al., 2007). Results of the GPC analysis are shown in Figure 8.6. The results of the softening point revealed differences both within the field samples and when

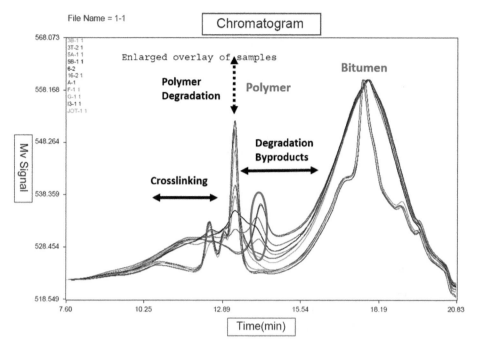

Figure 8.6 GPC chromatograms indicated differences in materials.

compared to both the new and manufacturer's specification for the waterproofing. In particular were differences between more liquid-like samples (ASTM, 2013; Lu & Isacsson, 1997; Lu et al., 1999) in the chromatogram as circled. Also, there was a general shift to smaller molecular weight (MW) molecules from chain scission. We also evaluated the ratio (Mw/Mn) of the Weight Average Molecular Weight (Mw) to the Number Average Molecular Weight (Mn). There were nominal differences proportional to the liquid-like aspects of the field samples, but not statistically significant.

The combined knowledge from these analyses was not sufficient to explain the causation; especially given that there was one subcontractor involved in the installation of the waterproofing and another for the snow melt system.

Continuing on with the investigation, small portions of the samples were ashed and the residue inspected by polarized light microscopy (PLM) to determine filler composition. The new material filler was consistent with carbonates, e.g., calcium or magnesium carbonate, and only a slight amount of other minerals (quartz/feldspar, gypsum, etc.) The field samples also had minor amounts of minerals (quartz) and cement/concrete components (likely debris from sample collection). We also evaluated some of the residue by scanning electron microscopy with energy dispersive X-ray (SEM-EDX) analysis and found residue consistent with magnesium and calcium carbonates as well as gypsum. The gypsum may be from a reaction of the calcium carbonate with the sulfur in the bitumen at elevated temperatures. Overall, this revealed varying amounts of carbonate filler compared to new product, supporting certain viscosity differences. Although the filler material was consistent with the material data sheet, in terms of type and amount, there was clearly less filler in the field samples than in the new material, which can affect creep resistance, where creep is the slow (months–years) deformation of a material under stress (Lackner et al., 2005).

8.3.4 Creep Testing

The manufacturer indicated that the water proofing had been tested independently at 60° C (140° F) for 5 hours (with no load) with <1 mm of deformation. A 3″ diameter, 3.5″ tall concrete core was placed on top of a heated subsample to simulate the installed and operating condition of the concrete over the waterproofing under heated conditions. The subsample was placed on the hot plate and kept at the following temperatures:

60° C (140° F), overnight, slight deformation
71° C (159° F), few hours, moderate deformation
84° C (183° F), 1 hour, severe deformation

We noted that the creep was mostly monolithic in nature, as the material spreads, but it did not flow. So, although the concrete slab over the waterproofing layer could cause some movement, it could not account for flowing condition.

8.3.5 Fabric Waterproofing Support

Small portions were also extracted from the field samples and the reinforcing fabric composition was determined by PLM. Samples were consistent with regard to reinforcement composition and placement according to manufacturer specifications. The sample collected around a drain was consistent with the manufacturer's special drain textile material. For the other samples, we observed a fabric mesh consisting of trilobal polyester fibers, consistent with the manufacturers reinforcing fabric. We also identified some polypropylene fibrous material. We observed that most of the apparent liquid-like oozing occurred at the felt layers. If the heating coils were a source of heat separation, one might expect the liquid to form closet to the coils (highest temperature); not farther away. The felt layers are not closest to the coils. Thus, an important spatial piece of information was acquired to help exclude the heating system as a likely source of liquification.

8.3.6 Initial FT-IR Analysis

Subsamples of the waterproofing field membranes were analyzed using Fourier transform infrared (FT-IR) microspectroscopy (Humecki, 1995; Roush, 1987). Subsamples were extracted from the main samples, put onto a low-e glass slide, then placed under a SensIR FT-IR microscope using an attenuated total reflection (ATR) objective to analyze the infrared reflected spectrum down to a 10-micrometer (um) approximate spot size. FT-IR spectroscopy has been widely used for bitumen- and asphalt-polymer matrices analysis (Blanco et al., 2001; Canto et al., 2006; Choquet, 1984; Daranga, 2005; Delpech et al., 2012; Helm & Petersen, 1968; Institute, 2010; Kawahara, 1969; Karlsson & Isacsson, 2003; Kawahara et al., 1974; Kunič et al., 2011; Lamontagne et al., 2001; Lu et al., 1999; Masson & Lacasse, 1999; Masson et al., 2001, 2002, 2003, 2006, 2011; Molenaar et al., 2010; Ruan et al., 2003; Wegan & Nielsen, 2000; Wu et al., 2009, 2010; Yang et al., 2010; Yu et al., 2011; Yut & Zofka, 2011).

All 5 field samples (including separation parts), new material, and new material heated under various time and temperature regimes were analyzed with FT-IR microspectroscopy. We made assumptions for peak assignment based on the literature review of the subject. These are presented in Table 8.1.

Table 8.1 FT-IR peak assignments.

Compound	Absorption Wave Number (cm^{-1})	Origin
Polystyrene (PS)	750	C-H Out of plane bending in monalkylated aromatics
	699	C-H Out of plane bending in monalkylated aromatics
Polybutadiene (PB)	993	C-H Out of plane bending of cis-alkene
	966	C-H Out of plane bending trans-alkene
	911	C-H Out of plane bending of terminal alkene
	730-650	C-H wagging of cis-alkene
Bitumen	1030	S=O Stretching
	874	C-H Out of plane bending
	814	C-H Out of plane bending
	746	C-H Out of plane bending
	722	Rocking of $(CH_2)_n$, $n > 4$
Butadiene Rubber (BR)	1450	CH of CH_2
Ozonolysis	1375	C=O stretch of ozonides

We then considered the polymer ratios as suggested by Masson (Masson et al., 2001):

966/699 for PB/PS
911/699 for PB/PS

We also considered the ratios (Choquet & Ista, 1992):

966/1376 for PB/Bitumen
698/1376 for PS/Bitumen

Since the heating trend in the control samples produced an increase in the PB/PS ratio, while the PB/Bitumen ratio remained fairly constant, this suggested that PB did not oxidize but that PS may have depolymerized (Masson et al., 2003) or was lost (e.g., by evaporation, leaching, etc.). The field samples, other than 6, also followed this trend (see Figure 8.7).

We noted two significant peaks at 1450 and 1375 cm^{-1} that developed only with high heating (500° F) over long periods (≥2 day). These are readily apparent in Figures 8.8 and 8.9 for the H-1, H-2, H-3, samples, and on the field samples. However, they are conspicuously absent or limited on other samples [A, B, C, D, E, F, G, I-1, I-2, and I-3] heated for shorter time periods (see Figure 8.10). This strongly indicates that the field sample material was overheated for long periods (>400° F). The 1450 cm^{-1} peak corresponds to a fairly strong SBR CH_2 bond, that may resist degradation more than the PS bonds. The 1375 cm^{-1} peak is indicative of an ozonide, suggesting a strong oxidation source (Yang et al., 2010).

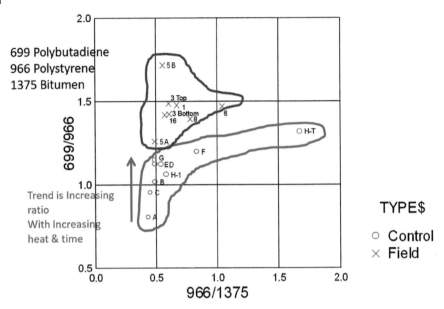

Figure 8.7 Polymer bitumen FT-IR Peak Ratios by sample type: Control (new untreated and heat treated in red) and Field samples (blue).

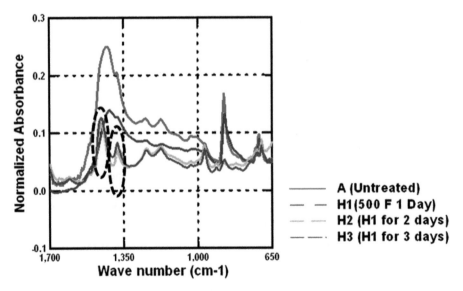

Figure 8.8 FT-IR spectra for new (unused) waterproofing before and after heat temperature treatment. Significant peaks at approximately 1450 and 1375 cm^{-1} developed only with high heating (500° F) over long periods (>=2 day).

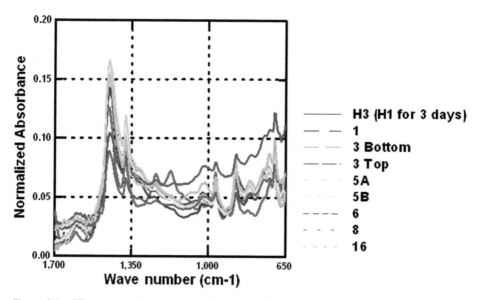

Figure 8.9 FT-IR spectra for new (unused) waterproofing after heat temperature treatment (H3) and all five field samples. Note peaks at approximately 1450 and 1375 cm^{-1} indicative of overheating over long periods.

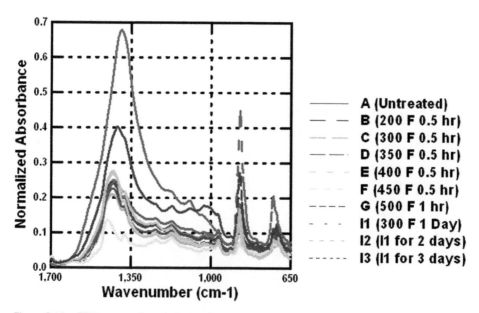

Figure 8.10 FT-IR spectra for new (unused) waterproofing before and after heat temperature treatment. Note peaks at approximately 1450 and 1375 cm^{-1} are limited or absent, especially compared to samples H2 & H3 that were heated longer at very high temperatures (shown in Figure 8.4).

8.3.7 Fluorescence Microscopy

A fluorescent microscope was used at 100-times to 1000-times magnification to evaluate primary fluorescence, in particular for the SBS polymer distribution. This ability to evaluate spatial distributions of components in samples is perhaps the microscope's greatest attribute. Fluorescence microscopy has been widely used for bitumen-polymer matrices analysis (Airey, 1999, 2004, 2011; Alonso et al., 2010; Choquet, 1984; Fawcett & McNally, 2000; Fawcett & McNally, 2001; Fernandes et al., 2008; García-Morales et al., 2004; Handle et al., 2016; Italia & Brandolese, 1996; Isacsson & Lu, 1999; Lu & Isacsson, 1997; Lu et al., 1999; Masson et al., 2011; Masson & Lacasse, 1999; Nevin et al., 2009; Oba et al., 1996; Polacco et al., 2011; Poulikakos & Partl, 2010; Qiu et al., 2012; Sawyer et al., 2008; Sengoz & Isikyakar, 2008; Soenen et al., 2006; Spadaro et al., 2011; Srivastava et al., 1992; Tapkin et al., 2011; Wegan & Nielsen, 2000). The selected use of fluorescence microscopy is based on the principle that polymers swell due to the absorption of some of the constituents of the base bitumen and due to the fluorescence effect in ultraviolet. The bitumen-rich phase appears dark or black, whereas the polymer- rich phase appears light (Sengoz & Isikyakar, 2008).

Fluorescence microscopy was used to evaluate all the field samples [1, 3 Top and Bottom, 5 (both parts A and B), 6, 8, and 16]. See Figure 8.11, a photomicrograph of top liquid-like material in sample 3T as a more typical example of the liquid-like material. To better assess the field samples, heating treatments on the new waterproofing were conducted and used for comparison. In Figure 8.12, sample A (untreated) shows compact pieces of SBS polymer (green-yellow) in a black bitumen matrix. Figure 8.13 is a new sample (heat treated at 500° F for 60 minutes). It reveals SBS polymer (green-yellow) expanding and redistributing in the black bitumen matrix. Also seen is a larger quasi-rhombohedral quasi-liquid component

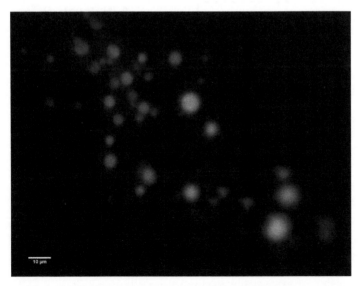

Figure 8.11 Photomicrograph of field sample 3 Top liquid-like material observed using fluorescence microscopy. Note the amount and distribution of small SBS polymer spheres in a matrix of bitumen.

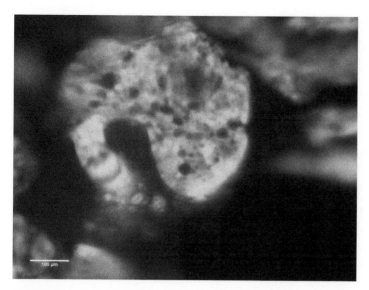

Figure 8.12 Photomicrograph of new sample A (untreated) observed using fluorescence microscopy. Note the compact pieces of SBS polymer (green-yellow) in a black bitumen matrix.

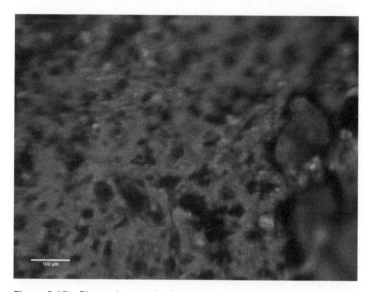

Figure 8.13 Photomicrograph of new sample G (heat treated at 500° F for 60 minutes) observed using fluorescence microscopy. Note the SBS polymer (green-yellow) expanding and redistributing in the black bitumen matrix. Also note the larger quasi-rhombohedral quasi-liquid component in the upper right beginning to separate (compare to Figure 8.14).

beginning to separate. Figure 8.14 shows a new sample heat treated at 500° F for 3 days. The SBS polymer (green-yellow) is also expanding and redistributing in the black bitumen matrix. But note the larger quasi-rectangular liquid component separated from the primary SBS polymer particles. This is believed to be correlated to the phase separation in the field samples. In general, it was observed that as the new waterproofing was heated

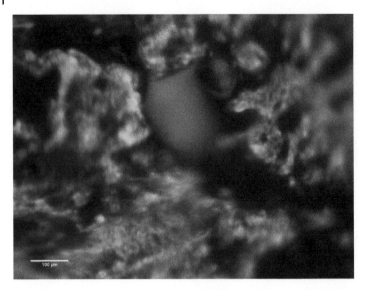

Figure 8.14 Photomicrograph of new sample I-3, observed using fluorescence microscopy. Again, the SBS polymer (green-yellow) is expanding and redistributing in the black bitumen matrix. Also note the larger quasi-rectangular liquid component separated from the primary SBS polymer particles.

at higher and longer temperatures, at first the SBS polymer dispersed and swelled. This was followed by restructuring and eventual phase separation consistent with that seen by others [LUX1997, LUX1999, MAS2003]. For the liquid-like phases in the field samples, SBS amounts were very low, poorly distributed, and spherical. Thus, although there was now strong data (FT-IR and fluorescence microscopy) to support that overheating was clearly evident and could cause phase separation, the lab heat treatment results did not completely explain the poorly distributed and spherical morphology of the SBS polymer.

8.3.8 Hot Stage and Fluorescence Microscopy

In order to understand the in-situ transformation observed in the treated samples as viewed by fluorescence microscopy, we used both an Instec HCS-400 hot-cold stage and a Mettler FP82 hot stage to evaluate the thermal varying characteristics of the microstructure of the samples. Microscopic size subsamples of material were evaluated using either the PLM or fluorescent microscope up to 400° C. Hot stage microscopy has been widely used for bitumen/polymer analysis [BAG2012, DEM2009, FRA2003, LAM2001, OBA1996, RAH1998, RAH2006], particularly in combination with fluorescence.

A new sample of the waterproofing was heated from room temperature to 300° C (572° F). The mixture softened at around 60° C (140° F) liquified at approximately 80° C (176° F), and showed some separation near 250° C (482° F). It was heated to 300° C (572° F) and lost material to volatilization as it was heated. Upon cooling (to room temperature), the polymer was less concentrated in volume. This helped explain the phase-separation process and the apparent loss of mass, but did not explain the spherical morphology and poor distribution.

The phase separation was supported by the heat-treat experiments; however, a quality assurance technician involved in the construction reported that Elephant Skin (a synthetic underlayment) was used in the construction. This material was observed in Sample 16, around a drain. In addition, it was stated that the material was melted prior to placing the full layer down. This product consists of a low density polyethylene (LDPE). Garcia-Morales [Gar2004] has found that LDPE–bitumen mixes begin to see separation of components by 140° C (284° F). Perez-Lepe (Pérez-Lepe et al., 2006, 2007) has found that LDPE–bitumen mixes have instability index values three-and-a-half orders of magnitude (approx. 3000 times) higher than SBS. If significant amounts of LDPE were melted and mixed with the waterproofing, the possibility for enhanced separation may have been a contributory factor to the oozing, but only at those locations where it was used. For the purposes of the investigation, this still pointed to the waterproofing contractor, and thus was not further considered. Although the pattern of the oozing did not support this cause as an only cause. In fact, spatially it would be limited in impact as it was only identified in one sample area.

8.3.9 Spherical Particle Formation Search

The search for this simple spherical morphology with poor and varied long-range correlative structure required another mechanism other than just a high temperature, extended heating regime. Our experience with spherical particles suggests two primary mechanistic actions: 1) an emulsion [a fine dispersion of minute droplets of one liquid in another in which it is not soluble or miscible] that was then solidified to some degree, or 2) a relatively quick cooling of a vapor (condensation).

One potential source of emulsification would be the application of a solvent. Solvents are often used to clean up bitumen and often used to help "cut" a waterproofing material in order to aid in lowering the viscosity. But not many solvents work well in creating a bitumen liquid phase. One solvent that dissolves bitumen well is xylene. We attempted to create an emulsion-like structure with the new waterproofing using silicon oil, naphtha, and xylene as solvents; as some solvents were suspected as having been applied to the waterproofing during installation (up to 104° C). None produced a spherical emulsion-like consistency. However, some phase separation was noticed with xylene (see Figure 8.15).

In terms of the second mechanism, quick vapor condensation often results in a spherical particle with Buckle providing initial conditions for metal-based spheritization (Buckle, 1978; Buckle & Mawella, 1985; Buckle et al., 1984; Buckle & Pointon, 1975, 1977; Buckle and Tsakiropoulos, 1981, 1985). The best modeling is associated with metal fume formation shown to be related to a Newtonian enthalpy model for undercooled solidification and departures from the Newtonian model adjusted for low Biot numbers (Levi & Mehrabian, 1982; Shukla et al., 2001). Although most data on sphere formation regards metals or silica-based compounds (like fly ash or fumed silica), our in-house air sampling studies have shown that plastic (including composites) may also create spherical particles in a similar manner.

In that vein, we conducted some experiments involving torching the new waterproofing. Torching resulted in both very little SBS polymer (like the field samples), and spherical distribution of the polymer. In Figure 8.16, sample H-1T (heat treated at 500° F for 8 hours then torched with a propane torch) demonstrates a low mass distribution of small SBS polymer spheres in a matrix of bitumen with some "hazy" areas consisting of a liquid

Figure 8.15 Photomicrograph of new sample A (unheated) treated with xylene observed using fluorescence microscopy. Note that the SBS polymer is still present in significant amounts in a "stabilized" form. However, there is a liquid phase revealed as hazy fluorescent areas surrounding the SBS.

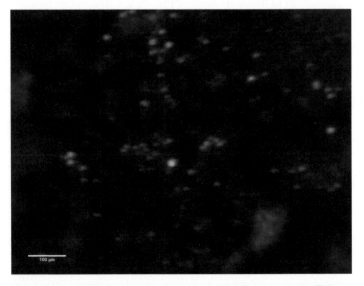

Figure 8.16 Photomicrograph of new waterproofing sample H-1T (heat treated at 500 F for 8 hours then torched with a propane torch) observed using fluorescence microscopy. Note the amount and distribution of small SBS polymer spheres in a matrix of bitumen with some "hazy" areas consisting of a liquid polymer phase.

polymer phase. This is consistent with the oozing material from the field. The presence of a fluid-like polymer phase that would be very leachable at room temperature could account for some of the SBS loss in the field samples. But one notes that torching at 800+° F also drives off the polymer resulting in a lot less SBS mass.

8.3.10 Discussion

The field material was visually and chemically consistent with an SBS-modified bitumen waterproofing. The filler composition of field samples was less than that of new samples but within the stated range. The general chemical composition of the waterproofing (SBS-modified bitumen) was revealed in the GPC, FT-IR, and fluorescence microscopy analyses.

The effect of heat at temperatures over 400° F was clearly indicated by control samples and the GPC, FT-IR, fluorescence microscopy, and hot stage microscopy results. Segregation occurs over 400° F, and more so for extended periods (≥=2 days). Chemical changes, indicative of overheating and extended heating, were revealed in the GPC shifts, the FT-IR peaks, and the polymer morphology and distribution by fluorescence microscopy; these are similar to those reported by Masson [MAS1999, MAS2008]. These characteristics were present in the field samples, indicating that they occurred during construction. The use of an unjacketed kettle, which is more difficult to maintain at the prescribed temperatures, was noted by one contractor. The laws of thermodynamics and heat transfer require that the temperature difference from the bottom to the top of a field kettle will be substantially greater than that in the lab due to increased height and diameter of the field compared to the lab container. Thus, the field temperatures at the bottom of the kettle (in order to get the top part liquid-like) would be substantially greater than the top. As a result, our lab simulation likely underestimated the overheating present in the field. In addition, because of the cold weather during which heating and application took place would only have created even more of a difference (from increased side wall losses to ambient temperature). Also, one worker indicated that he used an infrared thermometer on the top of the kettle and it registered a range of 450–475° F, which would make the bottom of the kettle much hotter still (≫475° F).

The fluorescence microscopy revealed that the only way to achieve both the very low concentration of SBS and the spherical (phase separation) morphology in the field samples was through a combined overheating and torching. However, either extended heating over 400° F or extended heating and torching could have resulted in phase separation and liquification (oozing). A post-analysis review of records and other interview sources from the construction indicated that torching or melting the top of the first pour surface occurred, consistent with the analyses and with the presence of oozing at the felt layers. The torching of the surface of the waterproofing was confirmed in a contractor letter indicating that "when the membrane has to be readjusted, ... the (waterproofing) will sometimes catch fire when using a Tiger Torch (weed burner)." A few photos of the use of a torch to soften the water proofing layer were also discovered.

This case required a considerable amount and range of microscopical analyses to solve the question of why the waterproofing layer failed (see Table 8.2). Not uncommon for complex cases, the final determination of failure required an evaluation using multiple tools and techniques as well as a good understanding of the products, materials, properties, and characterization options available. Microscopy was again the critical tool in finding the answer to a multimillion-dollar failure.

Table 8.2 Summary of analytical testing.

ID	Softening Pt.	GPC	u-FTIR	Filler Analysis	Fluorescence Microscopy	HSM	Creep
1	x	x	x	x	x		
3 Top		x	x	x	x		
3 Bottom		x	x		x		
5A	x	x	x	x	x		
5B	x	x	x		x		
6	x	x	x	x	x		
8	x	x	x		x		
16	x	x	x	x	x		
A	x	x	x	x	x	x	
B			x				
C			x		x		
D			x		x		
E	x		x		x		
F	x	x	x		x		
G	x	x	x		x		
H-1			x		x		
H-2			x		x		
H-3			x				
H-1T			x		x		
I-1			x				
I-2			x		x		
I-3	x	x	x		x		x
J-0T		x	x		x		

References

Airey, G. (2004). Styrene butadiene styrene polymer modification of road bitumens. *Journal of Materials Science, 39*(3), 951–959. https://doi.org/10.1023/B:JMSC.0000012927.00747.83

Airey, G. (2011). Factors affecting the rheology of polymer modified bitumen (PMB). In T. McNally (Ed.), *Polymer modified bitumen* (pp. 238–263). Woodhead publishing series in civil and structural engineering. Woodhead Publishing. ISBN 9780857090485. https://doi.org/10.1533/9780857093721.2.238

Airey, G. D. (1999). Dynamic shear rheometry, fluorescent microscopy, physical and chemical evaluation of Polymer modified Bitumens. *7th Conference on Asphalt Pavements for South Africa* (pp. 1–22). Nottingham, UK.

Alonso, S., Medina-Torres, L., Zitzumbo, R., & Avalos, F. (2010). Rheology of asphalt and styrene–butadiene blends. *Journal of Materials Science, 45*(10), 2591–2597. https://doi.org/10.1007/s10853-010-4230-0

ASTM. (2013). D7173-11, Determining the separation tendency of polymer from Polymer modified Asphalt. ASTM International, West Conshohocken, PA.

ASTM. (2013). D36/D36M-09, Standard test method for softening point of Bitumen (Ring-and-Ball Apparatus). ASTM International, West Conshohocken, PA.

Blanco, M., Maspoch, S., Villarroya, I., Peralta, X., González, J., & Torres, J. (2001). Determination of physico-chemical parameters for bitumens using near infrared spectroscopy. *Analytica Chimica Acta, 434*(1), 133–141. https://doi.org/10.1016/S0003-2670(01)00811-X

Buckle, E. (1978). Nature and origin of particles in condensation fume: A review. *Journal of Microscopy, 114*(2), 205–214. https://doi.org/10.1111/j.1365-2818.1978.tb00130.x

Buckle, E., & Mawella, K. (1985). Condensation and morphology of magnesium particles in vapour plumes. *Journal of Materials Science, 20*(7), 2647–2652. https://doi.org/10.1007/BF00556097

Buckle, E., Mawella, K., & Hitt, D. (1984). Particle condensation in metallic vapour plumes. *Journal of Materials Science, 19*(10), 3437–3442. https://doi.org/10.1007/BF00549836

Buckle, E., & Pointon, K. (1975). Condensation of cadmium aerosols. *Journal of Materials Science, 10*(3), 365–378. https://doi.org/10.1007/BF00543679

Buckle, E., & Pointon, K. (1977). Condensation of zinc aerosols. *Journal of Materials Science, 12*(1), 75–89. https://doi.org/10.1007/BF00738473

Buckle, E., & Tsakiropoulos, P. (1981). Solidification of metallic aerosol droplets from floating rafts: A test of the spiral growth mechanism for Cd and Zn. *Journal of Materials Science, 16*(4), 1103–1107. https://doi.org/10.1007/BF00542758

Buckle, E., & Tsakiropoulos, P. (1985). Condensation and crystal morphology of bismuth aerosols. *Journal of Materials Science, 20*(10), 3691–3696. https://doi.org/10.1007/BF01113777.

Canto, L., Mantovani, G., DeAzevedo, E., Bonagamba, T., Hage, E., & Pessan, L. (2006). Molecular characterization of styrene-butadiene-styrene block copolymers (SBS) by GPC, NMR, and FTIR. *Polymer Bulletin, 57*(4), 513–524. https://doi.org/10.1007/s00289-006-0577-4

Cash, C. G. (2004). *Roofing failures.* Routledge.

Choquet, F. (1984). Identification et proprietes de bitumes modifies ou non pour chapes d'etancheite. *Compte Rendu de Recherche,* (CR23/84). https://trid.trb.org/view/1032437.

Choquet, F. S., & Ista, E. J. (1992). The determination of SBS, EVA and APP polymers in modified bitumens. In *Polymer modified asphalt binders.* ASTM International.

Daranga, C. (2005). *Characterization of aged polymer modified asphalt cements for recycling purposes. Louisiana State University and Agricultural & Mechanical College.* (Order No. 3199724). Available from ProQuest Dissertations & Theses A&I. (304990006). Retrieved from http://unh-proxy01.newhaven.edu:2048/login?url=https://www.proquest.com/dissertations-theses/characterization-aged-polymer-modified-asphalt/docview/304990006/se-2

Delpech, M. C., Mello, I. L., Delgado, F. C., & Sousa, J. M. (2012). Evaluation of thermal and mechanical properties of rubber compositions based on SBR extended with safe oils. *Journal of Applied Polymer Science, 125*(5), 4074–4081. https://doi.org/10.1002/app.35560

Fawcett, A., & McNally, T. (2000). Blends of bitumen with various polyolefins. *Polymer, 41*(14), 5315–5326. https://doi.org/10.1016/S0032-3861(99)00733-8

Fawcett, A., & McNally, T. (2001). Blends of bitumen with polymers having a styrene component. *Polymer Engineering & Science, 41*(7), 1251–1264. https://doi.org/10.1002/pen.10826

Fernandes, M. R. S., Forte, M. M. C., & Leite, L. F. M. (2008). Rheological evaluation of polymer-modified asphalt binders. *Materials Research, 11*(3), 381–386. https://doi.org/10.1590/S1516-14392008000300024

Funkhouser, P. L. (1999). Reinforced waterproofing system for porous decks. *U.S. Patent No. 5,979,133*. Washington, DC: U.S. Patent and Trademark Office.

García-Morales, M., Partal, P., Navarro, F. J., Martínez-Boza, F., Mackley, M. R., & Gallegos, C. (2004). The rheology of recycled EVA/LDPE modified bitumen. *Rheologica Acta, 43*(5), 482–490. https://doi.org/10.1007/s00397-004-0385-4

Handle, F., Füssl, J., Neudl, S., Grossegger, D., Eberhardsteiner, L., Hofko, B., Hospodka, M., Blab, R., & Grothe, H. (2016). The bitumen microstructure: A fluorescent approach. *Materials and Structures, 49*(1), 167–180. https://doi.org/10.1617/s11527-014-0484-3

Helm, R., & Petersen, J. (1968). Compositional studies of an asphalt and its molecular distillation fractions by nuclear magnetic resonance and infrared spectrometry. *Analytical Chemistry, 40*(7), 1100–1103. https://doi.org/10.1021/ac60263a001

Humecki, H. J. (1995). *Practical guide to infrared microspectroscopy* CRC Press.

Institute, W. R. (2010). Asphalt Surface Aging Prediction (ASAP), Final Report. 124.

Isacsson, U., & Lu, X. (1999). Characterization of bitumens modified with SEBS, EVA and EBA polymers. *Journal of Materials Science, 34*(15), 3737–3745. https://doi.org/10.1023/A:1004636329521

Italia, P., & Brandolese, E. (1996). Polyproylene - Asphalt mixtures for water proofing membranes. *Symposium Modified Asphalts*, (pp. 1250–1253). Orlando, FL.

Karlsson, R., & Isacsson, U. (2003). Application of FTIR-ATR to characterization of bitumen rejuvenator diffusion. *Journal of Materials in Civil Engineering, 15*(2), 157–165. https://doi.org/10.1061/(ASCE)0899-1561(2003)15:2(157)

Kawahara, F., Santner, J., & Julian, E. (1974). Characterization of heavy residual fuel oils and asphalts by infrared spectrophotometry using statistical discriminant function analysis. *Analytical Chemistry, 46*(2), 266–273. https://doi.org/10.1021/ac60338a034

Kawahara, F. K. (1969). Identification and differentiation of heavy residual oil and asphalt pollutants in surface waters by comparative ratios of infrared absorbances. *Environmental Science & Technology, 3*(2), 150–153. https://doi.org/10.1021/es60025a002

Kunič, R., Orel, B., & Krainer, A. (2011). Assessment of the impact of accelerated aging on the service life of bituminous waterproofing sheets. *Journal of Materials in Civil Engineering, 23*(12), 1746–1754. https://doi.org/10.1061/(ASCE)MT.1943-5533.0000326

Lackner, R., Spiegl, M., Blab, R., & Eberhardsteiner, J. (2005). Is low-temperature creep of asphalt mastic independent of filler shape and mineralogy?—arguments from multiscale analysis. *Journal of Materials in Civil Engineering, 17*(5), 485–491. https://doi.org/10.1061/(ASCE)0899-1561(2005)17:5(485)

Lamontagne, J., Durrieu, F., Planche, J.-P., Mouillet, V., & Kister, J. (2001). Direct and continuous methodological approach to study the ageing of fossil organic material by infrared microspectrometry imaging: Application to polymer modified bitumen. *Analytica Chimica Acta, 444*(2), 241–250. https://doi.org/10.1016/S0003-2670(01)01235-1

Lee, S.-J., Amirkhanian, S. N., Shatanawi, K., & Kim, K. W. (2008). Short-term aging characterization of asphalt binders using gel permeation chromatography and selected Superpave binder tests. *Construction and Building Materials, 22*(11), 2220–2227. https://doi.org/10.1016/j.conbuildmat.2007.08.005

Levi, C., & Mehrabian, R. (1982). Heat flow during rapid solidification of undercooled metal droplets. *Metallurgical Transactions A, 13*(2), 221–234. https://doi.org/10.1007/BF02643312

Linde, S., & Johansson, U. (1992). Thermo-oxidative degradation of polymer modified bitumen. In *Polymer modified asphalt binders*. ASTM International.

Lu, X., & Isacsson, U. (1997). Compatibility and storage stability of styrene-butadiene-styrene copolymer modified bitumens. *Materials and Structures, 30*(10), 618–626. https://doi.org/10.1007/BF02486904

Lu, X., Isacsson, U., & Ekblad, J. (1999). Phase separation of SBS polymer modified bitumens. *Journal of Materials in Civil Engineering, 11*(1), 51–57. https://doi.org/10.1061/(ASCE)0899-1561(1999)11:1(51)

Masson, J., Collins, P., Robertson, G., Woods, J., & Margeson, J. (2003). Thermodynamics, phase diagrams, and stability of bitumen– polymer blends. *Energy & Fuels, 17*(3), 714–724. https://doi.org/10.1021/ef0202687

Masson, J., Collins, P., Woods, J., Bundalo-Perc, S., & Al-Qadi, I. (2011). Natural weathering of styrene–butadiene modified bitumen. In *Polymer modified bitumen* (pp. 298–335). Woodhead Publishing, Elsevier. https://doi.org/10.1533/9780857093721.2.298

Masson, J., Collins, P., Woods, J., Bundalopers, S., & Al-Qadi, I. (2006). Degradation of bituminous sealants due to extended heating before installation: A case study. In *85th Annual Meeting of the Transportation Research Board*, (pp. 22–26). Transportation Research Board.

Masson, J., & Lacasse, M. (1999). Effect of hot-air lance on crack sealant adhesion. *Journal of Transportation Engineering, 125*(4), 357–363. https://doi.org/10.1061/(ASCE)0733-947X(1999)125:4(357)

Masson, J.-F., Collins, P., Bundalo-Perc, S., Woods, J. R., & Al-Qadi, I. (2005). Variations in composition and rheology of bituminous crack sealants for pavement maintenance. *Transportation Research Record, 1933*(1), 107–112.

Masson, J.-F., Collins, P., Margeson, J., & Polomark, G. (2002). Analysis of bituminous crack sealants by physicochemical methods: Relationship to field performance. *Transportation Research Record, 1795*(1), 33–39.

Masson, J. F., Pelletier, L., & Collins, P. (2001). Rapid FTIR method for quantification of styrene-butadiene type copolymers in bitumen. *Journal of Applied Polymer Science, 79*(6), 1034–1041. https://doi.org/10.1002/1097-4628(20010207)79:6<1034::AID-APP60>3.0.CO;2-4

Masson, J.-F., Woods, J. R., Collins, P., & Al-Qadi, I. L. (2008). Accelerated aging of bituminous sealants: Small kettle aging. *International Journal of Pavement Engineering, 9*(5), 365–371. https://doi.org/10.1080/10298430802068899

Molenaar, A., Hagos, E., & Van de Ven, M. (2010). Effects of aging on the mechanical characteristics of bituminous binders in PAC. *Journal of Materials in Civil Engineering, 22*(8), 779–787. https://doi.org/10.1061/(ASCE)MT.1943-5533.0000021

Nevin, A., Comelli, D., Osticioli, I., Toniolo, L., Valentini, G., & Cubeddu, R. (2009). Assessment of the ageing of triterpenoid paint varnishes using fluorescence, Raman and FTIR spectroscopy. *Analytical and Bioanalytical Chemistry, 395*(7), 2139–2149. https://doi.org/10.1007/s00216-009-3005-4

Oba, K., Hean, S., & Björk, F. (1996). Study on seam performance of polymer-modified bituminous roofing membranes using T-peel test and microscopy. *Materials and Structures*, *29*(2), 105–115. https://doi.org/10.1007/s00216-009-3005-4

Pérez-Lepe, A., Martínez-Boza, F., Attané, P., & Gallegos, C. (2006). Destabilization mechanism of polyethylene-modified bitumen. *Journal of Applied Polymer Science*, *100*(1), 260–267. https://doi.org/10.1002/app.23091

Pérez-Lepe, A., Martínez-Boza, F., & Gallegos, C. (2007). High temperature stability of different polymer-modified bitumens: A rheological evaluation. *Journal of Applied Polymer Science*, *103*(2), 1166–1174. https://doi.org/10.1002/app.25336

Polacco, G., Fillipi, S., Markanday, S., Merusi, F., & Guilani, F. (2011). Fuel resistance of bituminous binder. In T. McNally (Ed.), *Polymer modified bitumen: Properties and characterisation* (pp. 336–365). Elsevier.

Poulikakos, L., & Partl, M. (2010). Investigation of porous asphalt microstructure using optical and electron microscopy. *Journal of Microscopy*, *240*(2), 145–154. https://doi.org/10.1111/j.1365-2818.2010.03388.x

Qiu, J., Van de Ven, M., Wu, S., Yu, J., & Molenaar, A. (2012). Evaluating self healing capability of bituminous mastics. *Experimental Mechanics*, *52*(8), 1163–1171. https://doi.org/10.1007/s11340-011-9573-1

Read, J., & Whiteoak, D. (2003). *The shell bitumen handbook*. Thomas Telford.

Roush, P. B. (1987). *The design, sample handling, and applications of infrared microscopes: A symposium*. ASTM International.

Ruan, Y., Davison, R. R., & Glover, C. J. (2003). Oxidation and viscosity hardening of polymer-modified asphalts. *Energy & Fuels*, *17*(4), 991–998. https://doi.org/10.1021/ef0202211

Sawyer, L., Grubb, D. T., & Meyers, G. F. (2008). *Polymer microscopy*. Springer Science & Business Media.

Sengoz, B., & Isikyakar, G. (2008). Analysis of styrene-butadiene-styrene polymer modified bitumen using fluorescent microscopy and conventional test methods. *Journal of Hazardous Materials*, *150*(2), 424–432. https://doi.org/10.1016/j.jhazmat.2007.04.122

Shukla, P., Mandal, R., & Ojha, S. (2001). Non-equilibrium solidification of undercooled droplets during atomization process. *Bulletin of Materials Science*, *24*(4), 547–554.

Snyder, L. R. (1969). Determination of asphalt molecular weight distributions by gel permeation chromatography. *Analytical Chemistry*, *41*(10), 1223–1227. https://doi.org/10.1021/ac60279a028

Soenen, H., Visscher, J. D., Vanelstraete, A., & Redelius, P. (2006). Influence of thermal history on rheological properties of various bitumen. *Rheologica Acta*, *45*(5), 729–739. https://doi.org/10.1007/s00397-005-0032-8

Spadaro, C., Plummer, C. J., & Månson, J.-A. E. (2011). Thermal and dynamic mechanical properties of blends of bitumen with metallocene catalyzed polyolefins. *Journal of Materials Science*, *46*(23), 7449–7458. https://doi.org/10.1007/s10853-011-5711-5

Srivastava, A., Hopman, P. C., & Molenaar, A. A. (1992). SBS polymer modified asphalt binder and its implications on overlay design. In *Polymer modified asphalt binders*. ASTM International.

Tapkin, S., Usar, U., Ozcan, S., & Cevik, A. (2011). Polypropylene fiber-reinforced bitumen. In T. McNally (Ed.), *Polymer modified bitumen: Properties and characterisation* (pp. 136–194). Elsevier.

Wang, Q., Liao, M., Wang, Y., & Ren, Y. (2007). Characterization of end-functionalized styrene–butadiene–styrene copolymers and their application in modified asphalt. *Journal of Applied Polymer Science, 103*(1), 8–16. https://doi.org/10.1002/app.23867

Wegan, V., & Nielsen, C. B. (2000). Microstructure of polymer modified binders in bituminous mixtures. In *Proceedings of the Papers Submitted For Review at 2nd Eurasphalt and Eurobitume Congress, Held* 20-22 September 2000, Euroasphalt and Eurobitume Congress, Barcelona, Spain. BOOK 1-SESSION 1.

Wu, S., Pang, L., Liu, G., & Zhu, J. (2010). Laboratory study on ultraviolet radiation aging of bitumen. *Journal of Materials in Civil Engineering, 22*(8), 767–772. https://doi.org/10.1061/(ASCE)MT.1943-5533.0000010

Wu, S.-p., Pang, L., Mo, L.-T., Chen, Y.-C., & Zhu, G.-J. (2009). Influence of aging on the evolution of structure, morphology and rheology of base and SBS modified bitumen. *Construction and Building Materials, 23*(2), 1005–1010. https://doi.org/10.1061/(ASCE)MT.1943-5533.0000010

Yang, B., Shi, Y., Fu, Z., Lu, Y., & Zhang, L. (2010). A study of the ozonolysis of butadiene rubber in the presence of ethanol. *Polymer Degradation and Stability, 95*(5), 852–858. https://doi.org/10.1016/j.polymdegradstab.2010.01.010

Yu, J.-Y., Feng, Z.-G., & Zhang, H.-L. (2011). Ageing of polymer modified bitumen (PMB). In T. McNally (Ed.), *Polymer modified bitumen: Properties and characterisation* (pp. 264–297). Woodhead Publishing Series in Civil and Structural Engineering. Woodhead Publishing. ISBN 9780857090485. https://doi.org/10.1533/9780857093721.2.264.

Yut, I., & Zofka, A. (2011). Attenuated total reflection (ATR) Fourier transform infrared (FT-IR) spectroscopy of oxidized polymer-modified bitumens. *Applied Spectroscopy, 65*(7), 765–770.

9

Value of Multiple Associations

Introduction

The value of multiple associations is based on the ability to corroborate a conclusion, thus demonstrating its truth or validity. Although multiple associations do not remove all doubts or alternative explanations, it does provide support or strength for what has already been suggested. This is used in all areas of reasoning and scientific inquiries. The reader may be most familiar with the common use of multiple associations in diagnosing medical diseases. An experienced physician is able to determine the likely ailment of a patient by putting together multiple different symptoms. For example, at the time of this writing, the world is suffering from the COVID-19 pandemic. Initial diagnoses of COVID-19 is based on individuals self-reporting a variety of symptoms (i.e., headaches, fever, cough, chills, shortness of breath or difficulty breathing, fatigue, muscle or body aches, headache, new loss of taste or smell, sore throat, congestion or runny nose, nausea or vomiting, diarrhea), which on their own could be due to a large range of medical or environmental factors. However, when multiple symptoms are experienced together, it lends weight to the conclusion that a person has this disease. Further, the rarer the symptom, the stronger its indication of the disease. Thus, a person with a headache may not think they are positive for the coronavirus; however, someone who suddenly experiences a loss of taste or smell would more likely reason that they have this illness. And if a person wakes to find that they have both a headache and a loss of taste or smell, the suspicion that they have COVID would be reinforced. There is meaningful value added to a conclusion when it is based on multiple associations, even if no statistical analysis can be used to provide quantitative support. Although there is still the possibility that several links could be entirely coincidental, a larger number of associations reduces this possibility.

Microscopes are excellent tools for observing and interpreting multiple associations. Microscopes are able to analyze a large range of materials with assorted properties and mixtures, thus enabling it to outperform other scientific instruments. This chapter provides several case examples to showcase the ability of a microscopist to use one or several microscopes to make multiple associations which are then used to solve diverse problems.

Solving Problems with Microscopy: Real-life Examples in Forensic, Life and Chemical Sciences, First Edition.
Edited by John A. Reffner and Brooke W. Kammrath.
© 2024 John Wiley & Sons Ltd. Published 2024 by John Wiley & Sons Ltd.

9.1 Atlanta Child Murders Investigation

John A. Reffner, Ph.D.[1] *and Brooke W. Kammrath, Ph.D.*[2,3]

[1] *John Jay College of Criminal Justice, New York, NY, USA*
[2] *University of New Haven, West Haven, CT, USA*
[3] *Henry C. Lee Institute of Forensic Science, West Haven, CT, USA*

Between July 1979 and May 1981, at least 28 African American children, adolescents, and adults were murdered in Georgia, in what would become known as the Atlanta Child murders. The first two victims, Edward "Teddy" Smith and Alfred Evans, were boys ages 14 and 13, respectively, who disappeared four days apart but were found on July 28th in a wooded area. The first was shot in the upper back with a .22-caliber weapon while the other died from strangulation. Two more boys, Milton Harvey and Yusef Bell, went missing in early September and October; their bodies were recovered in November in remote or abandoned areas. At first, law enforcement did not associate these crimes together due to the different causes of death and disposals of the bodies. The remains of 13 more children were found in 1980, and another 7 boys and 6 adult males were recovered in 1981. A task force was created to catch this serial murderer, which included approximately a dozen stakeouts of bridges along the Chattahoochee River. Wayne Williams was first identified as a suspect on May 22nd, 1981, when detectives beneath a bridge heard a splash while other detectives saw a light-colored 1970 Chevrolet station wagon drive over the bridge. The driver of the station wagon would eventually be arrested, tried, and found guilty of two of the murders, and associated to many more through fiber evidence. This case captured the attention of the nation, and showed the world how microscopy could be used to link multiple victims to each other and with the serial killer, Wayne Williams.

The authors do not intend to write a treatise on this case, and instead will highlight the essential role an expert microscopist played in the discovery and analysis of the physical evidence that was critical in the investigation. For readers interested in learning about other details of the case, there are numerous books, articles, television shows, and documentaries of varying quality, depth, and neutrality. We recommend a 2010 CNN documentary titled "Atlanta Child Murders" which gives a comprehensive and balanced account of the events and evidence.

On May 19th, 1980 the body of the seventh victim, 14-year old Eric Middlebrooks, was discovered. When examining the crime scene, Detective Robert Buffington observed a tuft of fibers caught in one of the victim's tennis shoes, at the edge of the rubber sole (Figure 9.1). He hypothesized that the fibers were transferred to the shoe when the victim was dragged across a carpeted floor. Detective Buffington brought this evidence, along with other traces, to criminalist Larry Peterson at the Georgia State Crime Laboratory. Peterson had been a forensic examiner for several years prior to this case, and knew the potential value fiber evidence could have in an investigation. This brought the fiber evidence to the forefront of this investigation.

At that time, the role of a trace evidence examiner was multifaceted. In addition to laboratory examination of traces by microscopy and chemical instrumentation, forensic scientists also went to crime scenes, autopsies, and other locations of interest where evidence was collected. In this case, it was the careful examination of the victim's bodies and clothing at autopsy as well as the suspect's residence and vehicles by the scientific experts which enabled the recognition and identification of probative fibers and hairs. It is unfortunate that the trend today is to exclude the scientist from this role and instead have evidence discovery, documentation, and collection performed by nonscientist officers.

(A)

(B)

Figure 9.1 A crime scene photograph of Eric Middlebrook's tennis shoes, with the red tuft of fibers circled (A), and a bright field photomicrograph of the red fibers (B), obtained from screenshots of the CNN documentary "Atlanta Child Murders."

At the trial of Wayne Williams for the murders of Jimmy Ray Payne and Nathaniel Cater in January 1982, Peterson testified about his examinations of the remains of these two victims, as well as the suspect's residence and vehicle. When examining the body and red shorts of Jimmy Ray Payne on April 27th, 1981, Peterson collected traces of fibers and hairs, fingernail scrapings, and known exemplars of head and pubic hairs from the victim. On May 24th, 1981, Peterson went to the Chattahoochee River near the I-285 Bridge, to examine the crime scene where the nude body of Nathaniel Carter, which was secured to a boat, was discovered. Peterson also went to the Fulton County Medical Examiner's office to perform a careful examination of the body, where he collected a number of fibers and hair material from the mud-caked head hair of the victim. Additionally, Peterson removed a questioned hair from the victim's pubic hair and exemplar hairs from the victim. Next, Peterson examined Williams' vehicle, a light beige Chevrolet Concourse station wagon, at the FBI's garage building on June 3, 1981. And last, he went to Williams' residence in Atlanta on June 3rd with FBI special agent Hal Deadman, and again on June 22nd where he collected samples of carpet and other textile materials as well as hairs from the family dog.

Numerous different types of fiber and hair evidence were found on many of the victims. This evidence was used to link the victims together, and validated the theory that there was a serial killer in Atlanta preying on young African Americans. Table 9.1 shows the identified sources of the fiber evidence associated with 14 of the victims.

Table 9.1 This table represents some but not all of the items from Williams' home, vehicles, or person that contained fibers that were also recovered from these victims (Deadman, 1984a, 1984b).

Name of victim	Violet & Green Bedspread Williams Bedroom	Green Carpet Williams Bedroom	Dog Hairs Williams Bedroom	Yellow Blanket Williams Bedroom	Blue Rayon Fibers Debris from Williams Home	Trunk Liner 1978 Plymouth	Carpet 1970 Ford	Carpet 1970 Chevrolet	Kitchen Carpet	Ford Truck Liner	Backroom Carpet	Porch Bedspread	Glove jaket	YellowNylon	White Polyester
Alfred Evans	X	X	X			X									
Eric Middlebrooks	X	X	X				X			X				X	
Charles Stephens	X	X	X		X					X	X			X	X
Lubie Geter	X	X	X					X	X		X				
Terry Pue	X	X	X												X
Patrick Baltazar	X	X	X	X				X					X	X	X
Joseph Bell	X				X										
Larry Rogers	X	X	X	X				X				X		X	
John Porter	X	X	X	X	X			X				X			
Jimmy Payne	X	X	X	X	X			X							
William Barrett	X	X	X	X	X			X					X		
Nathaniel Cater	X	X	X	X							X				

The yellow-green ("English Olive") Wellman nylon 6,6 carpet fibers that were in Wayne Williams' bedroom and also recovered from 17 of the victims was the most probative evidence in this case. This trilobal fiber had a unique cross section, with two large lobes and one short one (Figure 9.2). Prior to identifying Williams as a suspect, the investigators thought that finding the manufacturer of these unique fibers could provide an investigative lead in this case. As such, the story about finding the fiber manufacturer is one that deserves sharing. Peterson and Deadman sent photomicrographs of the fiber to several fiber experts and contacts within the textile industry, all of whom agreed that they had

(A)

(B)

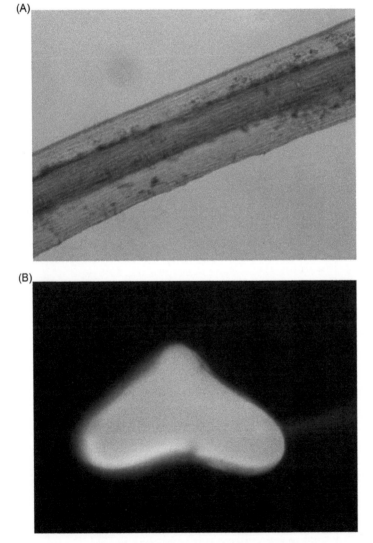

Figure 9.2 Photomicrographs of an evidence fiber recovered from one of the victims of the Atlanta Child Murders, showing a longitudinal view in plane-polarized light (A) and a cross section in cross-polarized light (B). This nylon 6,6 carpet fiber, identified as being a Wellman 181B fiber dyed "English Olive," has a unique cross-sectional shape.

never encountered a fiber with this cross-sectional shape. Fortuitously, a meeting of chemists at a fiber manufacturer's research facility occurred around this time. Over lunch, one forensic chemist shared photomicrographs of the yellow-green fiber with a woman who was in charge of examining the cross-sectional morphology of competitor's carpets with the aim of protecting copyrights. Upon seeing a photomicrograph of the cross section, she was able to identify the fiber as being a Wellman 181B fiber. Many consider this to be a lucky coincidence, while we maintain that it is consistent with diligence and the adage "fortune favors the prepared mind," which is discussed in Chapter 6. Once the fiber's manufacturer was identified, a detailed analysis of its evidentiary significance in these murders was performed, which was the first time this was done with fiber evidence. This is described in an article by Hal Deadman (1984b) entitled "Fiber Evidence and the Wayne Williams Trial (Conclusion)," and includes an extensive probability determination analysis. It was concluded that "the chance of randomly selecting an occupied housing unit in metropolitan Atlanta and finding a house with a room having carpet like Williams' carpet was determined to be 1 chance in 7,792 – a very low chance" (Deadman, 1984b). Some important details that were used in this analysis include:

- the fact that Wellman sold this carpet, the "Luxaire" line, to West Point Pepperell Company in Georgia, who only sold it for a period of 1 year (December 1970 to December 1971).
- no other manufacturer was identified as producing a fiber with the same cross-sectional shape.
- West Point dyed the carpet 16 different colors, and one-sixteenth of the small production line was dyed the color "English Olive."
- West Point Pepperell sold 5710 square yards of this fiber to 10 southeastern states which included Georgia.

The quality and depth of the information obtained regarding the Wellman 181B fiber's distribution is not often obtainable to forensic scientists. Although the unique cross-sectional shape provides value for the association in terms of its identification, the calculation of statistical probabilities regarding its prevalence in the population support its significance.

There was controversial fiber evidence in this case, specifically concerning a yellow fiber that was recovered on several of the victims. FBI special agent Hal Deadman collected a fiber sample of a yellow blanket found under the bed of Williams' during the June 3rd search, but did not seize the blanket. By the second search on June 22nd, the blanket had disappeared. Fibers from the blanket were shown to be consistent with those recovered from the victims, but the source had vanished. While some hypothesize that Williams got rid of the blanket, Williams claims that the yellow blanket never existed and the evidence was fabricated. Although it is not uncommon for evidence at a crime scene to be sampled rather than seized, the blanket should have been documented through photography. This is a minor oversight, but it has become an item of interest for conspiracy theorists.

There were several other recovered fibers of significance. Twenty-one victims had an unusual violet acetate fiber which was interwoven with green cotton. This fiber was consistent with fibers collected from Williams' bedspread, which was composed of these two fibers woven together. Dog hairs that were similar to those of the Williams' family dog, a

German shepherd named Sheba, were recovered from 20 of the victims. Over the course of the killings, Williams drove three different cars, and fibers from each were recovered from 10 victims sequentially. First, Williams drove a Plymouth, and fibers from that car were found on Alfred Evans. When Williams got rid of that car, those fibers never showed up on later victims. Williams' family next bought a Red Ford LTD, and fibers from that vehicle were found on victims in the middle of the timeline. Last, in October 1980, Williams' purchased a Chevrolet Station wagon, and fibers from that car were recovered from six of the later victims.

Forensic fiber comparisons are not often straightforward because there may be changes to a fiber when it is exposed to the elements. Anyone who has left a textile in the sun for a period of time or over-washed a garment will understand that fiber colors fade. In this case, the last few victims' remains were recovered from the Chattahoochee River, and it was evident that the bodies had been in the water for a few days. When Peterson compared the evidence yellow-green fibers to the exemplar carpet fibers from Williams' bedroom, he observed that the evidence fibers had a slightly duller color. This was a problem because they did not match. There was thus a need to determine if this was a meaningful difference (see Chapter 10) or an explainable difference. Peterson hypothesized that the color differences in the fibers were caused by extended exposure to the river water, and designed an experiment to test this. Peterson put some of the known fibers from the bedroom into water collected from the Chattahoochee River, and after a few days he observed the bleaching of the fibers by the water. This resulted in their faded appearance which was comparable to the evidence fibers collected from the bodies. Peterson followed the scientific method, and as a result was able to explain the differences in a meaningful way which corresponded with known events.

Approximately 30 years after the murders, mitochondrial DNA testing was performed on the human hairs recovered from the body of Patrick Baltazar. Results of the mitochondrial DNA testing, which is used to determine a common maternal ancestor and as a result is not as discriminating as nuclear DNA analysis, showed that Williams was a possible source of the hairs. In 2010, special agent Hal Deadman, who was now the director of the FBI's DNA laboratory, stated in the CNN documentary (Polk, 2010), "It would probably exclude 98% or so of the people in the world," which is the strongest finding possible with this particular type of testing. Of 1148 African-American hair samples in the FBI's mitochondrial DNA database, only 29 had the same sequence, which calculates to only 2½ of every 100 African-Americans. Further, none of the Caucasian or Hispanic hair samples in the database had the same mitochondrial DNA profile, which together makes the odds almost 130-to-1 against the hairs coming from someone other than Williams (McLaughlin, 2019). This mitochondrial DNA result adds to the multiple associations made between Williams and the victims.

The public has been fascinated by this Atlanta child serial murder case for more than 40 years, and it continues to be the subject of great attention. It is interesting to note that once Wayne Williams was arrested, these child murders stopped. This case demonstrates the value of multiple associations; ultimately the hair and fiber evidence in this case is compelling. When presented with the totality of the evidence, the jurors convicted Wayne Williams of the two murders after 11 hours of deliberation. In an interview with one of the jurors, he said, "There were just too many fibers placed on too many bodies.... What would the chances be of finding these same, all of these fibers...? The chances would be just astronomical" (Polk, 2010). Williams was ultimately sentenced to two consecutive life terms, and remains incarcerated despite several attempted appeals.

References

Deadman, H. (1984a). Fiber evidence and the Wayne Williams trial (Part I). FBI Law Enforcement Bulletin, pp. 13–20.

Deadman, H. (1984b). Fiber evidence and the Wayne Williams Trial (Conclusion). FBI Law Enforcement Bulletin, pp. 10–19.

McLaughlin, E. C. (2019). *Unsolved Atlanta child murders are back under the microscope.* Cable News Network (CNN). Retrieved from https://www.cnn.com/2019/03/21/us/atlanta-child-murders-wayne-williams-mayor-bottoms/index.html

Polk, J., (Senior Producer). (2010). *The Atlanta Child Murders* [TV documentary]. Cable News Network (CNN) Documentary.

9.2 Hog Trail Murders

John A. Reffner, Ph.D.[1] and Brooke W. Kammrath, Ph.D.[2,3]

[1] John Jay College of Criminal Justice, New York, NY, USA
[2] University of New Haven, West Haven, CT, USA
[3] Henry C. Lee Institute of Forensic Science, West Haven, CT, USA

In the mid-1990s, the bodies of five men who had been raped, murdered, and mutilated were found within a 10-mile radius of the woods in Charlotte County, Florida. Wild boars roamed this area of remote woods, which were aptly nicknamed the Hog Trails. The first set of remains recovered were that of a still-unidentified man who was found in Punta Gorda, Florida, on February 1, 1994. He had signs of rope burns on his skin and the medical examiner determined that his body had been left outside for approximately a month. On January 1, 1996, the second set of remains was recovered in North Port, Florida, when a dog brought a male human skull home from the woods. Investigators were able to find most of the skeleton, and it was determined that his genitals had been removed, but he was never identified. The third victim, eventually identified as John William Melaragno, was recovered in North Port on March 7, 1996. The fourth and fifth victims, Kenneth Lee Smith (who was identified by a tattoo on his shoulder and confirmed through dental records) and Richard Allen Montgomery, respectively, were found in the woods on April 17, 1996, after Smith's skull was discovered by a hiker. It was evident that a serial killer was operating in this area, and a concerted investigation commenced.

A suspect, Daniel Conahan Jr., quickly emerged. He was first identified by David Allen Payton, an inmate at Glades Correctional Institute, who claimed that Conahan had picked him up on the side of the road on March 5, 1995, presenting him with alcohol and drugs. Payton stated that Conahan offered to pay him $100 to pose for nude pictures in the woods, which he declined. However, Payton became suspicious of Conahan, and was able to escape by driving away in the car. Payton was arrested later that day, charged with grand theft auto for steeling Conahan's father's vehicle, a Blue 1984 Mercury Capri. Later, two different men who had been propositioned, Charles Bateman and Robert Beckwith, went to the police and provided sketches of the perpetrator. Although they were not able to identify Conahan from a 3-year-old driver's license photograph, Batemen identified the blue Capri when brought by police to an area near Cohahan's home. Last, in June 1996, the murders were connected to the kidnapping and attempted murder of Stanley Burden in 1994. Mr. Burden, a male prostitute, reported to the Fort Myers police department that he had been propositioned, tied to a tree, assaulted, and nearly strangled by a man named "Dan" who drove a gray Plymouth station wagon. This event was supported by hospital records. When Burden was shown a photo array line-up by police, he identified Conahan as the perpetrator.

In July 1996, police obtained a warrant to search and seize evidence from Conahan's home, his Plymouth station wagon, and his father's Mercury Capri. There was abundant physical evidence associating Conahan with the murder of Montgomery, which was the only charge levied against the defendant at trial. There were credibility issues with the witnesses, so the physical evidence was critical in this case. Most significant was a multi-layered automotive paint chip which was recovered from the pubic hair of Montgomery and determined to be analytically indistinguishable from Conahan's father's blue Mercury Capri. Next was the fiber evidence shown in Table 9.2. There were 17 different types of

Table 9.2 This table represents the 17 different types of fibers recovered from the crime scene which had associations with Conahan's vehicles or home. Those in red font were associated with a known source while those in red font have an unknown provenance.

		Blue Red Nylon Split Film	Red Nylon (Plymouth Carpet)	Pink Nylon (Mercury Carpet)	Pink Polyprop ylene (rope)	Purple Brown Acetate	Gold & Black Acrylic	Yellow Rayon	Green Acrylic	Red Black Cotton	Red Nylon (Plymouth Seat Fabric)	Tan Acrylic	Black Cotton	Black Acrylic	Black Polyester	Blue Nylon (tarp)	Blue Polyester (back Pack)	Green Wool
Defendants vehicles	Plymouth	Potential Sorce						7	14		Potential Sorce							
	Mercury			Potential Sorce	2	6	32	8	17	66		1	1					1
	Bedroom					11	3	395	<102	100s		1		1	2	Potential Sorce	Potential Sorce	
Scence	"M" Body	1	1	2														
	Carpet Padding "M" Body				Potential Sorce	1	1											
	Body Parts							2	18	1	1							
	Coat					397			1	7		Potential Sorce	Potential Sorce	Potential Sorce	Potential Sorce	1	1	
	Rope "B" Assault																	4

fibers recovered from the Montgomery crime scene and body which were also found at the defendant's home and/or in his vehicles. This evidence included fibers from the defendant's gloves which were associated with those found on a tree, and a pink polypropylene fiber which was microscopically indistinguishable from a piece of rope found in Conahan's father's car.

Although there were multiple fiber associations, similar to the Atlanta Child Murder case detailed above, this case presented a nuanced problem. Unlike the Atlanta case, many of the associated fibers did not have a known source of the originating textile. The comparison fibers were recovered from locations the defendant frequented (i.e., his home and cars) but their exact source remains unknown. For example, green acrylic fibers were recovered from Montgomery's mutilated body parts, as well as Conahan's bedroom, his Plymouth, and his father's Mercury; however, no textile was recovered containing this fiber. Not having an identified source reduces the probative value of those fibers, but does not eliminate the validity of association. The significance of the fiber evidence in this case was further bolstered by the unusual nature of some of the fibers, such as acetate fibers with a purple-brown color and the gold and black acrylic fibers.

In August 1999, Conahan was convicted of the first-degree premeditated murder and kidnapping of Richard Allen Montgomery. A sixth body was found in 1997, and since the trial 12 additional bodies and skeletons have been recovered in the same area and under the same circumstances as the initial five victims, suggesting Conahan may have killed at least 18 young men. To date, Conahan continues to proclaim his innocence despite the strong evidentiary significance of the paint and multiple fiber associations.

Acknowledgments

We gratefully acknowledge Prof. Peter R. De Forest for providing important details of this case, including the information contained in Table 9.2.

9.3   Hoeplinger Murder

Henry C. Lee, Ph.D.[1,2]*, John A. Reffner, Ph.D.*[3] *and Brooke W. Kammrath*[1,2]

[1] University of New Haven, West Haven, CT, USA
[2] Henry C. Lee Institute of Forensic Science, West Haven, CT, USA
[3] John Jay College of Criminal Justice, New York, NY, USA

On May 7, 1982, at 4:55 a.m., a 911 police dispatcher in an affluent neighborhood in Easton, Connecticut, received a phone call from a nearly hysterical man. Mr. John C. Hoeplinger called 911 that morning to request an ambulance come to his home to treat his wife Eileen who he described as having blood "all over her face and head" and "in serious shape." Officer Raymond Osborne was the first to arrive to the scene at 5:01 a.m., and Mr. Hoeplinger met him at the front door sobbing, distraught, and shaken. Mr. Hoeplinger was wearing a white undershirt and blue jeans, both containing bloodstains, and had a cut on a finger of his left hand. Officer Osborne was led into the family room where he observed the body of Mrs. Eileen Hoeplinger. She was lying face up on the couch, wrapped in a sheet and blankets, with blood covering her face and head, and her skull appearing to be partially caved in. It was clear to this first responding officer that Mrs. Hoeplinger was dead, and he escorted Mr. Hoeplinger outside to wait in a police cruiser. When additional police officers and medical personnel arrived, the two young Hoeplinger children, ages 5 and 2.5, who were sleeping upstairs, were transported to a neighbor's house.

Mr. Hoeplinger described the events of the early morning to a second police officer and again later in a 16-page narrative to the chief of police while at the Easton police department. He stated that he awoke in the middle of the night to go to the bathroom, and saw that his wife was not with him. He looked into the bedrooms of his two children and other rooms in the house, to no avail. He then searched areas outside of his home. While looking around his front driveway, Mr. Hoeplinger volunteered that he saw his wife covered in blood in a wooded area. He described that he embraced his wife, and then went to the house to get a sheet and blankets which he wrapped her in before carrying her into the house. He placed her onto the living room couch. Mr. Hoeplinger also stated that he tried to clean up the blood from the downstairs areas so as not to frighten his children should they come downstairs. Then he called 911.

The crime scene presented several challenges to the initial police investigators, specifically with interpretation of the blood traces. There was a bloody trail on the winding gravel driveway with blood on both sides of it (Figure 9.3). These complex blood configurations were uninterpretable to initial police investigators, so they called for the assistance of Dr. Henry C. Lee, the Director of the Connecticut State Forensic Science Laboratory and an accomplished Forensic Scientist with a specialization in crime scene reconstruction, among other things. The blood traces in the driveway told Dr. Lee a different story than that detailed by Mr. Hoeplinger. Dr. Lee saw separate blood trails – some parallel and some intersecting. He examined the path of blood, inch-by-inch, to reveal the movements of the body on the driveway. The blood trail began at the house, then led about 25 ft from the front doorway. At that location, there was an accumulation of the bloodstains and noted signs of hesitation. The pool of blood and the size of other blood droplets at this location indicated that the perpetrator had stopped here with the bleeding body and put the body closer to

(A)

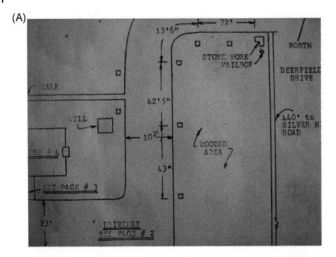

(B)

(C)

Figure 9.3 Crime scene diagram (A) and photographs of the driveway (B–E) showing the blood trail.

(D)

(E)

Figure 9.3 (Cont'd)

the ground. The blood trail continued into the woods and then lead back out of the woods to the driveway where it ended at a collection of blood smears. The evidence indicated that the body was dropped at this location and subsequently dragged further away from the house. The direction of the dragged body was indicated by the crushed and broken leaves and vegetation which were all bent in the same direction away from the house. Further, blood was only found on one side of the leaves. This blood trail ended at another point of hesitation, and then the trail which consisted of drag marks and more blood traces led into a second area of the woods where a large pool of blood, some hair, and tissue material were discovered. The victim was then carried back into the house. Ultimately, there were two separate trails of blood: one traveling away from the house and another going into it.

Mr. Hoeplinger self-identified himself as carrying his wife from the woods to the house; however, the perpetrator of the crime and the source of the movements away from the house remained in question. This blood evidence provided a detailed reconstruction of the events; however, it remained uncertain as to whether it was an unknown assailant or Mr. Hoeplinger who had bludgeoned Mrs. Hoeplinger and then carried and dragged the body along the driveway and into the woods.

The living room also contained valuable blood evidence for reconstruction (Figure 9.4). Police investigators initially observed only a large pool of blood on the sofa, which was where the body was placed by Mr. Hoeplinger, and was attributed to post-event bleeding. However, the quantity of blood indicated to Dr. Lee that this was the location of the bludgeoning. Additionally, the coffee table in front of family room couch had approximately 50 individual small blood droplets (1–2 mm in diameter), interpreted by Dr. Lee as a cast off

(A)

(B)

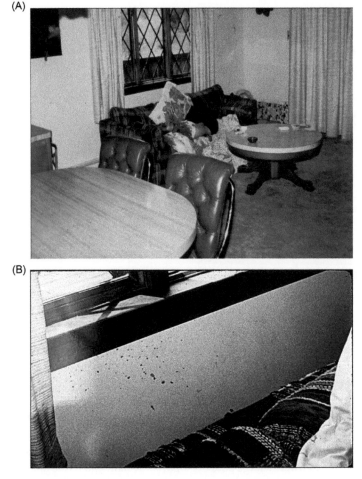

Figure 9.4 Photographs of the crime scene living room, overall (A) and mid-range of the blood traces on the wall behind the couch.

pattern. The presence of these blood traces was not explainable by just carrying the body as Mr. Hoeplinger had claimed. Dr. Lee hypothesized that Mrs. Hoeplinger was bludgeoned while lying on the couch and hit 3-to-4 times with the brick. Cast off from the latter blows caused the cast off observed on the coffee table. Additional blood traces found on walls, window, and drapes behind the couch further supported the conclusion that the bludgeoning occurred inside the family room.

The Medical Examiner quickly determined that Mrs. Hoeplinger died from strangulation and deep cranial trauma as a result of being hit on the head with a blunt object which resulted in massive bleeding. It was important to identify the weapon used in the bludgeoning. Police investigators initially thought the blunt object was a missing golf club, specifically a 7-iron. They thought this because they found in the basement a full set of golf clubs that was missing the 7-iron. Police called in the dive team to search the private lake behind the house for the golf club, as it provided a potentially easy place for the perpetrator to dispose of the weapon. However, the divers came up empty handed, and no club was recovered from the lake.

Dr. Lee was not a golfer, but thought the blood patterns were inconsistent with having come from a golf club. Further, the injuries of the victim showed a square corner, which was also not consistent with a golf club. He suggested they look for a small heavy object with right angles. Dr. Lee thought the weapon could have been a brick from the driveway or one from a newly built wishing well. When Dr. Lee told the divers about his brick theory, one of the divers recalled seeing some blood on some of the shallow rocks at the edge of the lake and a brick in this same area. The brick was not initially collected because they were looking for a golf club. The divers collected the brick and it was brought to the forensic laboratory for microscopic analysis of trace transfers on the surface.

The initial stereomicroscopic examination of the brick was very informative (Figure 9.5). One portion of the brick was shown to have direct contact with a human head, due to the presence of blood, human tissue, and crushed head hairs. The head hairs from the brick were microscopically compared to those collected from the body of Mrs. Hoeplinger, and were found to be consistent with sharing a common source. In addition, the fact that portions of the hairs were crushed indicated that this was indeed the murder weapon. Further, the Medical Examiner identified the brick as a possible murder weapon due to its capacity to be used to cause the severe head trauma that was documented at autopsy.

The last item of important physical evidence was a wet white t-shirt found hanging on the back porch (Figure 9.6 and 9.7). The shirt contained diluted blood traces, thus was immediately of interest. The shirt was sent to the forensic laboratory for a microscopical examination. In addition to the small blood traces, when samples were taken and examined with a brightfield microscope, green algae and diatoms were found. Subsequent analysis with a scanning electron microscope was able to identify the specific types of algae and diatoms, which were consistent with that found in the lake and recovered on the brick, which had been submerged in the lake (Figure 9.8). This led to the conclusion that whoever had worn this t-shirt was the person who placed the brick in the pond. Auspiciously, the t-shirt was the same size and make as that being worn by Mr. Hoeplinger at the time of the event and more of these were found in his wardrobe. It was this microscopic evidence which convincingly indicated that Mr. Hoeplinger had murdered his wife.

(A)

(B)

(C)

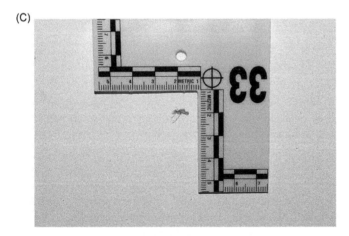

Figure 9.5 Stereomicroscope photomicrographs of the brick, overall (A) and with increased magnification to better visualize the crushed hairs and human tissue (B), which were removed for subsequent analysis (C).

Figure 9.6   Photograph of the back porch and pond.

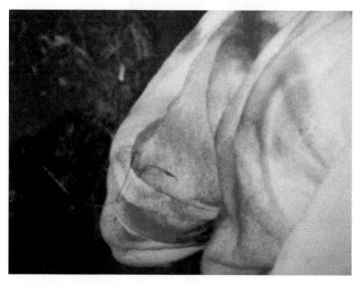

Figure 9.7   Photograph of the recovered t-shirt from the back porch.

In conclusion, the physical evidence refuted Mr. Hoeplinger's statement regarding an intruder being the perpetrator. Although the blood evidence was critical for reconstructing the events and proving the "how," it was the microscopic evidence which was used to identify both the weapon and the murderer. Mr. Hoeplinger was found guilty of the crime of manslaughter in the first degree by a jury, and sentenced to a term of imprisonment of 20 years (State of Connecticut v. John C. Hoeplinger, 1988). Although this verdict was reversed on appeal due to legal issues, Mr. Hoeplinger was convicted in a second trial and died in prison (State of Connecticut v. John C. Hoeplinger, 1992a and 1992b).

(A)

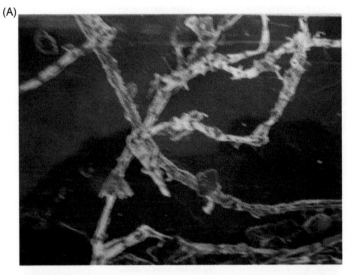

(B)

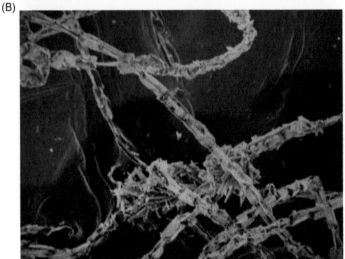

Figure 9.8 Scanning electron photomicrographs of the diatoms and algae; known samples taken from the pond (A) and from the t-shirt sample (B).

References

Hoeplinger & Madison/Hawaii (Season 1 Episode 1)[TV series episode]. (2004, June 2). In
  Trace Evidence: The Case Files of Dr. Henry Lee.
State of Connecticut v. John C. Hoeplinger. 206 Conn. 278 (1988).
State of Connecticut v. John C. Hoeplinger. (1992a). 609 A.2d 1015 (Conn. App. 1992), 9791.
State of Connecticut v. John C. Hoeplinger. (1992b). 27 Conn. App. 643 (1992).

9.4 Jackson Pollock Authentication

John A. Reffner, Ph.D.[1] and Brooke W. Kammrath, Ph.D.[2,3]

[1] John Jay College of Criminal Justice, New York, NY, USA
[2] University of New Haven, West Haven, CT, USA
[3] Henry C. Lee Institute of Forensic Science, West Haven, CT, USA

The authentication of the what is believed to be the last work of Jackson Pollock, titled "Red, Black and Silver" (Figure 9.9), has been the subject of considerable controversy in the art world for over 60 years (https://www.thedailybeast.com/

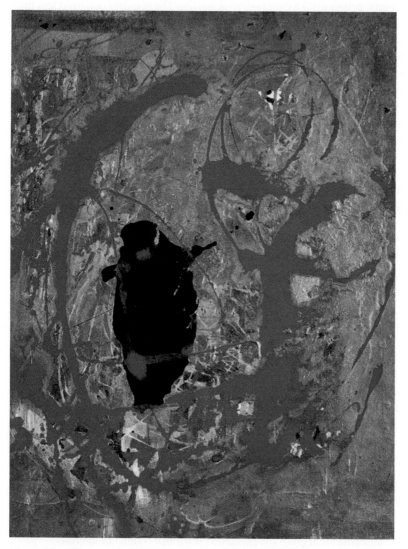

Figure 9.9 Photograph of the controversial Jackson Pollock painting "Red, Black and Silver," taken by Nicholas Petraco.

red-black-and-silver-just-may-be-jackson-pollocks-last-painting; https://www.today.com/news/new-evidence-offered-authenticity-pollocks-purported-final-work-wbna53505945; https://arthive.com/jacksonpollock/works/378650~Red_black_and_silver; Kligman, 1974). The conflict began with two women: Lee Krasner, Jackson Pollock's widow, and Ruth Kligman, his young mistress. While Kligman claimed that the painting was a "love letter" to her which Pollock made a few weeks prior to his death in 1956, Krasner maintained that Pollock had not painted in over 2 years prior to his death due to his alcoholism (Cohen, 2013).

Ruth Kligman detailed her relationship with Pollock and the painting of "Red, Black and Silver" in her 1974 biography "Love Affair: A Memoir of Jackson Pollock" (Kligman, 1974). According to Kligman, Krasner discovered the affair and sailed to Europe during the summer of 1956. Kligman then moved into Pollock's home in East Hampton, New York, where the two lived together until a car crash took his life. The accident occurred less than a mile from their home, and while Pollack and another passenger, Edith Metzger, were killed in the crash, Kligman survived. However, a few weeks prior to his death, Kligman, an aspiring artist, asked Pollock to teach her how to paint. She brought Pollock a canvas, some paint, and sticks while he was outside on his lawn. According to Kligman, when he had finished the painting, Pollock stated "Here's your painting, your very own Pollock."

Lee Krasner was an accomplished abstract expressionist artist prior to her relationship with Pollock. Upon his death, she immediately returned from her summer European vacation with friends to manage his estate in New York. Krasner disliked Kligman based on their personal history, and Krasner always insisted that "Red, Black and Silver" was a crude fake that was not made by Pollock.

Upon Pollock's death, Lee Krasner became the sole executor of his estate. A few years later Krasner advocated for cataloging Pollock's large volume of work. Eugene V. Thaw, a private art dealer and close friend of Krasner, recalled discussions about this while he and his wife visited the Pollock-Krasner home in the 1960s. While at Springs, Thaw volunteered to lead the project, and together with Francis V. O'Connor, an art historian, they published the five-volume set "Jackson Pollock: A Catalogue Raisonné of Paintings, Drawings and Other Works" in 1978 (https://www.catalogueraisonne.org/profile/pollock). This catalog did not include "Red, Black and Silver," because its authenticity was contested by Krasner and the authors of the collection. Upon Krasner's death in 1984, she bequeathed $23 million to establish the Pollock-Krasner Foundation to provide financial support for promising artists. Additionally, from 1990 through 1996, the Foundation supported the Pollock-Krasner Authentication Board, which was tasked with authenticating and cataloging Pollock's works.

The authentication process of "Red, Black and Silver" has been historically fraught with bias. Kligman and others believed that the personal animosities between her and Krasner influenced the opinions of the art connoisseurs (Kligman, 1996; Rosenberg, 1965), which ultimately resulted in the painting not being certified. Traditionally, art connoisseurs examine the composition and brush strokes of a piece to determine if a specific artist had created it. Since this is a subjective process which is based on the training and experience of the connoisseur, it is possible that the friendship between Krasner, O'Connor, and Thaw biased their authentication determination which in turn, influenced all proceeding ones. Although Krasner and her colleagues had intimate knowledge of Pollock and his work, there remains questions about the validity of this authentication given the previously described conflict of interests.

On the other side of the dispute, Kligman had the support and testimony from three of the most prominent, experienced, and well-respected figures in American art, all of whom endorsed the painting's authenticity (Ashton, 1994; Frank, 1994; O'Connor, 1995). Further, in 2011, Phillips de Pury and Company (Phillips) hired Orion Analytics to conduct an independent scientific investigation regarding the authenticity of "Red, Black and Silver" (Martin, 2011). Orion Analytical is a micro-niche materials analysis and consulting firm with expertise in paint, art, and cultural heritage analysis, which in 2016 was acquired by Sotheby's to become its worldwide Department of Scientific Research. James Martin, the founder of Orion Analytical, is a well-recognized expert in paint analysis and cultural heritage investigations, and is the Senior Vice President and Director of Scientific Research at Sotheby's. Martin's expertise in authentication has included the analysis of other Pollock paintings and has previously been solicited by the Pollock Authentication Board. Martin's examination of "Red, Black and Silver" included a macroscopic visual inspection using bright white light and long-wave UV light, infrared photography (to detect organic matter), stereomicroscopy, Fourier transform infrared (FT-IR) microspectroscopy, and confocal dispersive Raman microspectroscopy. In his final report on "Red, Black and Silver" (Untitled) Martin wrote the following conclusion: "If the claims of Ms. Kligman and her friend Bette Waldo Benedict are true, accurate, and complete, then Untitled is a work which Jackson Pollock painted in 1956" (Castelli, 1994).

In 2012, Nicholas Petraco initiated a new microscopical approach to the authentication of "Red, Black and Silver." Petraco is an accomplished microscopist and former forensic scientist with the New York City Police Department. Among other skills, Petraco specialized in trace evidence analysis, which included the microscopic detection and identification of paint, hairs, fibers, dust, plants, minerals, and many more minute materials. With over 40 years of experience identifying microscopic evidence, Petraco was uniquely qualified to answer this authentication problem. The fundamental question was determining whether the painting was made at the East Hampton home of Pollock or in Kligman's Manhattan studio (see Figures 9.10 and 9.11).

Petraco's first examination of the Pollock painting occurred at the Phillips Auction House in New York City on September 5th, 2012. At this point, he documented the painting

Figure 9.10 A photograph of the East Hampton home of Pollock, provided by Nicholas Petraco.

as received, and observed several items underneath the paint, which included possible hairs, seeds, and other miscellaneous traces (Figure 9.12). These observations prompted Petraco to request further investigation of the painting in a laboratory setting.

On November 19th, 2012, "Red, Black and Silver" was brought to the John Jay College of Criminal Justice Criminalistics Laboratory for an in-depth scientific examination (Figure 9.13). The first step of the scientific interrogation of the painting was to unwrap it and remove the frame. It was then documented and examined with a variety of illuminations, including the use of polarized light, ultraviolet light and infrared photography. The ultraviolet light examination revealed the presence of a possible biological stain, but no further analysis was performed to investigate this sample (Figure 9.14).

Having discovered potential trace evidential material on "Red, Black and Silver" which could help prove where the painting was crafted, Petraco next needed a source of knowns to compare

Initial Observation of Painting at Phillips Auction House on 9/5/2012

Appears to be possible hairs, fibers, seeds, and so on underneath the paint

Figure 9.12 A composite of the close-up photograph of possible potential trace evidence under the paint of Pollock's "Red, Black and Silver," as observed and documented by Nicholas Petraco at Phillips auction house (Nicholas Petraco).

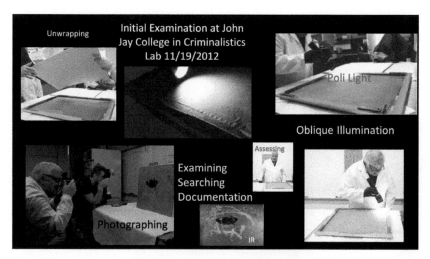

Figure 9.13 An overview of Petraco's initial examination of "Red, Black and Silver" at his laboratory in John Jay College's Criminalistics Laboratory on November 19th, 2012.

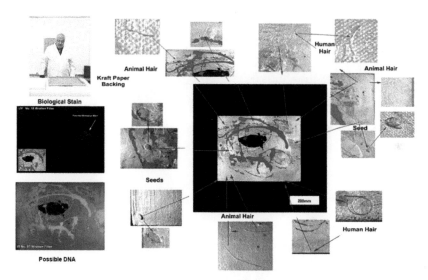

Figure 9.14 A review of the preliminary results of Petraco's initial examination of "Red, Black and Silver" at his laboratory in John Jay College's Criminalistics Laboratory on November 19th, 2012.

to the questioned traces. Thus, Petraco requested access to the two locations where the painting was alleged to have been painted: Kligman's Manhattan studio and Pollack's home and studio in East Hampton. The latter is now designated as the Pollock/Krasner National Historic Site owned by Stonybrook University. Permission was granted for access to both places. On February, 15, 2013, Petraco arrived at Pollock's home and studio located at 830 Springs, Fireplace Road, Easthampton, NY, where he was met by the director of the historic site.

Upon arrival at the site, Petraco and his team were escorted to the attic located in Pollock's home and given various items once owned by Pollock to examine, as illustrated in Figures 9.15 and 9.16. All possible traces were collected from Pollock's personal items, safeguarded and packaged for subsequent examination and comparison.

Figure 9.15 Photograph of Petraco and his assistant in Pollock's attic preparing to examine items owned by Jackson Pollock.

(A)

(B)

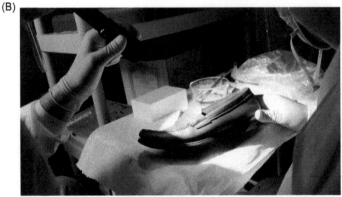

Figure 9.16 Photographs of Petraco and his assistant examining Jackson Pollock's shoes in the attic of the Pollock/Krasner National Historic Site (A & B). Potential DNA traces were sampled via swabbing (C), and hair, fiber, and other traces (D) were collected for subsequent analysis.

(C)

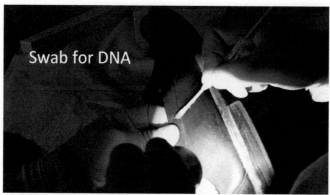

(D)

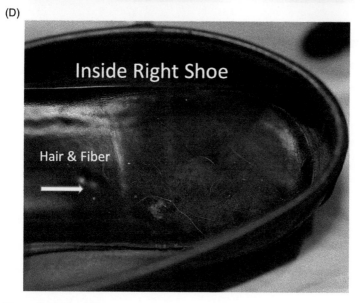

Figure 9.16 (Cont'd)

Next, the trace materials collected from Pollock's personal property were examined in Petraco's laboratory. All the sealed packages of trace materials were opened, carefully sorted, documented, and examined with stereomicroscopy. Next, all the items were mounted and initially identified using polarized light microscopy (PLM). Figure 9.17A-B illustrates the comprehensive stereomicroscopy and PLM examinations that were conducted by Petraco.

Now that Petraco had materials from Pollock's home which could be compared to the materials embedded in "Red, Black and Silver," the painting was returned to the John Jay College Criminalistics Laboratory on June 24, 2013 for the removal of a selected number of specimens embedded underneath the layers of paint, as shown in Figure 9.18. In total, eight questioned specimens were removed from "Red, Black and Silver" for examination and comparison purposes, designed P1-P8, marked for identification, and packaged for future examination (Figure 9.19).

A

B

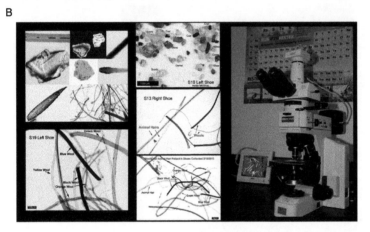

Figure 9.17 The traces collected from Pollack's property were each first examined with a stereomicroscope, sorted and documented, then prepared for PLM analysis (A). A composite image illustrating some of the items recovered from Pollock's shoes and subsequently identified with a PLM are shown in B.

After Petraco had examined questioned samples from Pollock's personal items and questioned samples from the painting, he requested that he be allowed return to the Pollock-Krasner House and Study Center to collect known specimens from the grounds. Permission was granted and on July 11, 2013 Petraco went to East Hampton New York to collect known specimens from the site. Specimens of soil and vegetation were collected from outdoor areas, and known rug specimen were collected from Pollock's attic. Figure 9.21A-B depict the known specimens collected at this site.

Petraco now had in his possession eight questioned specimen from the painting, samples from Pollack's clothing and residence, and known standards. Over the course of the next month he conducted numerous microscopical examinations and comparisons of the questioned and known specimens. First, he conducted an examination and comparison of the brown human head hairs from the painting with the brown human head hairs from Pollock's shoes. Unfortunately, known head hair specimen from Pollock's head were unavailable for comparison. Petraco used brightfield microscopy to identify two types of

(A)

(B)

Figure 9.18 "Red, Black and Silver" was removed by a professional art conservator at John Jay College and prepared for the removal of the specimens (A), then examined by Petraco as shown in (B) where he is removing a questioned hair specimen, designated as P1, from the painting on June 24, 2013.

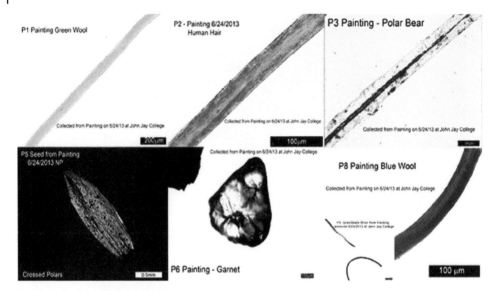

Figure 9.19 A composite photograph of each questioned specimen, designated as P1-P8, removed from the painting in question.

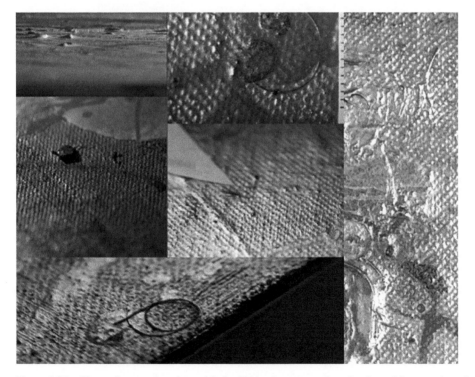

Figure 9.20 Photomicrographs taken with the PLM microscope of a selection of the questioned materials recovered from "Red, Black and Silver" on June 24, 2013.

Figure 9.21 Known soil and vegetation specimens collected from Jackson Pollock's yard, now the Pollock-Krasner House and Study Center (A), and known dyed wool and polar bear hair specimens collected from Jackson Pollock's attic acquired on July 11, 2013.

light brown colored human head hairs in Pollock's shoes. Additionally, there were two questioned light brown colored head hairs embedded in the painting which, when examined with a comparison brightfield transmission microscope, were shown to be consistent in color to those found in Pollock's shoes (Figure 9.22).

Next, Petraco used a brightfield microscope to compare the questioned polar bear hair fragment (designated as P3) he found embedded in "Red, Black and Silver" and polar bear hairs recovered from Pollock's shoe (Q6-S13) with known polar bear hair standards from the attic in Pollock's home and known polar bear hair standards he had obtained from the United States Customs Laboratory (Figure 9.23). Petraco concluded that P3 was indeed

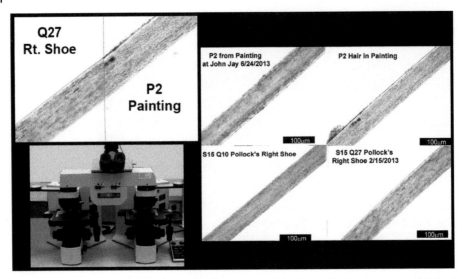

Figure 9.22 Human head hair comparison of the two questioned light brown colored human head hairs found in RBS and the numerous light brown colored head hairs obtained from Pollock's shoes.

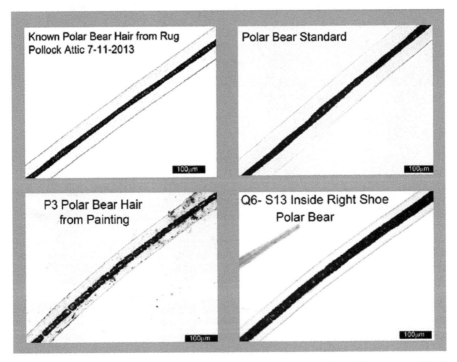

Figure 9.23 Photomicrographs of the questioned and known polar bear hairs compared by Petraco.

a fragment of hair from a polar bear and that it could have originated from the rug in Pollock's attic.

Petraco then conducted a comparison of the garnet grain he recovered from the painting (P6) and the grains of garnet he collected from inside Pollock's right shoe (S15 Q29c) with the known garnet grains he recovered from the known soil specimens collected in Pollock's yard. First Petraco used PLM to identify all garnet grains as being Almandine garnet. Next, he utilized X-ray diffraction (XRD) to confirm his findings (Figure 9.24A). Finally, Petraco used micro X-Ray fluorescence (XRF) spectroscopy to determine the elemental composition the garnets. Micro-XRF demonstrated that the garnet mineral grain recovered from the painting was analytically indistinguishable from an almandine garnet sample collected from the soil outside of Pollack's East Hampton home and studio as well as a garnet sample collected from Pollack's right shoe (Figure 9.24B).

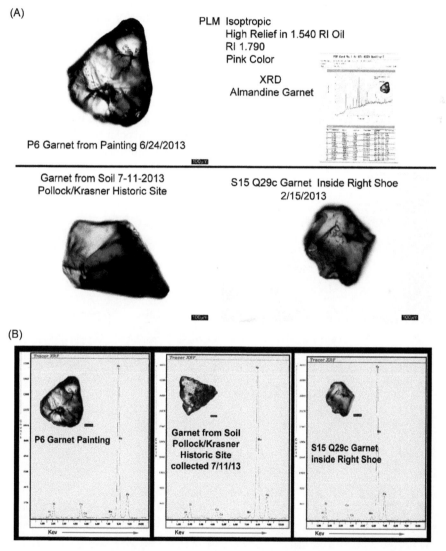

Figure 9.24   XRD identification and comparison of the questioned and known almandine garnet specimens (A), and micro-XRF spectra of mineral grains of garnet, together with their photomicrographs, collected from the "Red, Black and Silver" painting (left), soil from Pollack's East Hampton home and studio (middle), and inside Pollack's right shoe (right).

Next Petraco used a stereomicroscope, brightfield microscope, and PLM to identify and compare the seeds he found in Pollock's left shoe (Q13 – S19) and in "Red, Black and Silver" (sample P5) with the grass seeds from Pollock's yard. All seed samples were identified as American beachgrass, which is a species of grass native to eastern North America and found on the east coast of the United States from Maine to North Carolina (Figure 9.25).

Last, with the aid of a visible light microspectrophotometer, Petraco compared the spectra from the two different colored wool fibers (P1 and P8) found embedded in the painting, to those collected from the wool fibers found in Pollock's shoes and those from the multicolored wool rugs stored in Pollock's attic (Figures 9.26 and 9.27). He concluded that the dyes used in the multicolored wool rugs stored in Pollock's attic were the same dyes found in the wool fibers from Pollock's shoes and the wool fibers found embedded in "Red, Black and Silver." Therefore, the wool rugs in Pollock's attic could be the source of the wool fibers in his shoes and in the painting.

This case demonstrates the value of multiple associations for solving problems of provenance. It is important to recognize that value comes not only from the number of associations, but the uniqueness of each association. In this case, the associations made between the minerals and fibers are convincing, but not as valuable as the polar bear hair which is relatively rare to find in daily life. To further elucidate the significance of these multiple associations, statistical analysis was performed by Prof. Nicholas D.K. Petraco (John Jay

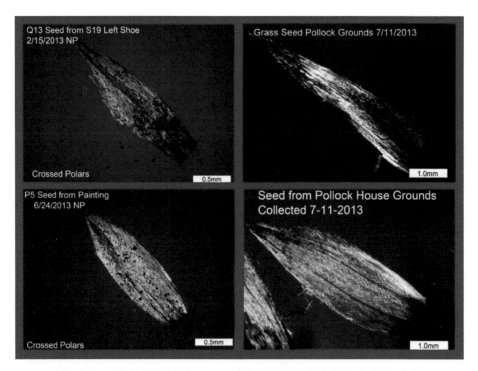

Figure 9.25 Composite image of PLM photomicrographs, using crossed polarized light, of seeds collected from the left shoe found in Pollock's attic (top left), the "Red, Black and Silver" painting (bottom left), and Pollack's house grounds (top and bottom right).

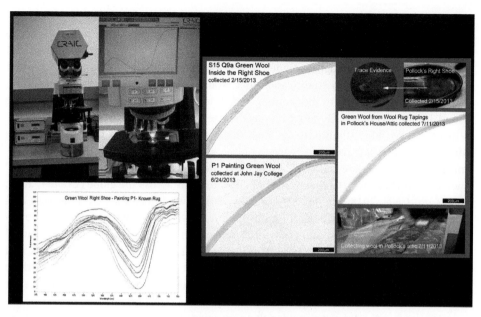

Figure 9.26 Composite image of PLM photomicrographs of all the green colored wool fibers collected from inside Pollock's right shoe (top left), the "Red, Black and Silver" (P1) painting (bottom left), and a rug in Pollock's attic (middle right). Photographs of the shoe and rugs.

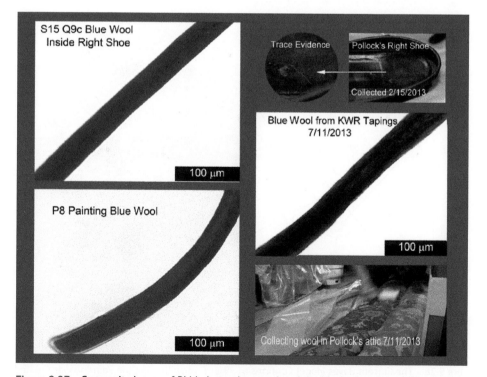

Figure 9.27 Composite image of PLM photomicrographs of the blue colored wool fibers collected from inside Pollock's right shoe (top left), the "Red, Black and Silver" painting (bottom left), and a rug in Pollock's attic (middle right). Photographs of the shoe and rugs are also depicted.

College of Criminal Justice). Prof. Petraco performed a Bayes Network analysis, which is used by NASA for modeling and monitoring critical systems performance. Prof. Petraco made a Bayes network model for the question: Did Jackson Pollack paint "Red, Black and Silver"? The statistical model concluded to a 99+% probability that this work was created at Pollack's East Hampton studio, and most probably by Jackson Pollack (Figure 9.28). Although the science proves the painting was indeed painted at Pollock's home on Long Island, it cannot conclusively prove he was the one who put the brush to canvas.

In the art world, authenticity of "Red, Black and Silver" is still not resolved, thus continuing the controversy between connoisseurship and science. Although present-day news is replete with stories of "fake but, expertly authenticated paintings" owned by world class art galleries and major museums across the globe (Cooper, 2016; Fox, Kara, CNN, 2018; Stuble, 2018; Salaky, 2019), world-class art connoisseurs still inflexibly maintain that "true authorship cannot be established without an expert evaluation of the composition and individual strokes that reveal an artist's "signature." (Cohen, 2013) In spite of the overwhelming scientific evidence that "Red, Black and Silver" is a genuine Pollock painted in July 1956 just weeks before his tragic death as Ruth Kligman described in her sworn testimony, it appears that the decision to reject the authenticity of this painting was based only on the traditional methods of connoisseurship, and sadly did not consider the new scientific evidence. In the future, these authors are hopeful that the contributions of science, and in this case the associative microscopic traces, will eventually be appreciated and recognized for the value they bring to authentications of works of art.

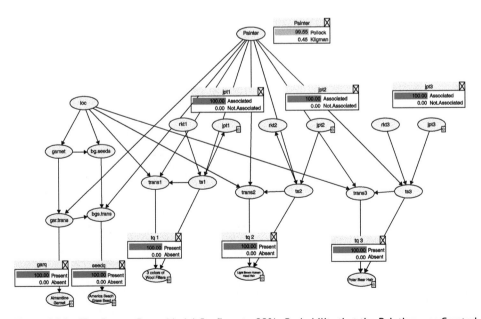

Figure 9.28 The Current Bayes Model Confirms to 99%+ Probability that the Painting was Created at Springs, and most likely by Jackson Pollock.

Acknowledgment

We gratefully acknowledge Mr. Nicholas Petraco, Natalie Zayne, and Professor Nicholas D.K. Petraco for providing important details about this case as well as the images and photomicrographs.

References

Ashton, D. (1994, December 1). Art critic for *The New York Times*, Professor of art history at Cooper Union University and a senior critic in painting and printmaking at Yale, wrote in a letter addressed to Francis O'Connor and the PKAB. I have no reason to doubt the authenticity of this painting, which, it seems to me, is utterly characteristic of Pollock.

Castelli, L. (1994, December 1). Wrote in a letter addressed to Francis O'Connor and the PKAB. To the best of my knowledge and belief, this painting entitled *Red, Black and Silver* is by Jackson Pollock.

Cohen, P. (2013, November 24). *A Real Pollock? On This, Art and Science Collide*. New York Times. https://www.nytimes.com/2013/11/25/arts/design/a-real-pollock-on-this-art-and-science-collide.html

Cooper, A. (2016, May 22). *$80 Million Con: 60 MINUTES*. The biggest, most lucrative art fraud scam in U.S. history, where a prestigious old New York art gallery sold fake works for 15 years. https://www.cbsnews.com/news/60-minutes-80-million-art-fraud-anderson-cooper

Fox, Kara, CNN. (2018, April 30). *French museum discovers more than half of its paintings are fakes*. CNN. https://www.cnn.com/style/article/french-museum-half-paintings-fake-intl/index.html

Frank, E. (1994, December 3). A Pulitzer-prize winning scholar and Pollock biographer, wrote in a letter addressed to Francis O'Connor and the PKAB. My eye tells me that in all probability the painting is by Pollock, and I believe in actual fact it is.

https://www.catalogueraisonne.org/profile/pollock.

https://www.thedailybeast.com/red-black-and-silver-just-may-be-jackson-pollocks-last-painting

https://www.today.com/news/new-evidence-offered-authenticity-pollocks-purported-final-work-wbna53505945

Kligman, R. (1974). *Love Affair: A Memoir of Jackson Pollock*.

Kligman, R. (1996, December 3). In an affidavit dated.

Martin, J. (2011, May 11). Orion analytical, orion project no. 1688, *Report Untitled (Estate of Ruth Kligman)*, p. 17.

O'Connor, F. (1995). Introduction. *JCPR Supplement One*, xv.

Rosenberg, H. (1965, January). The art establishment. *Esquire*.

Rosenberg, H. Archived January 14, 2012, at the Wayback machine. National Portrait Gallery, Smithsonian Institution. Retrieved February 22, 2008.

Salaky, K. (2019, January 31). *10 times famous works of art turned out to be fake*. INSIDER. https://www.insider.com/cases-of-faked-and-forged-artwork-2019-1

Stuble, P. (2018, April 28). *Art gallery discovers more than half of its paintings are fake*. INDEPENDENT. https://www.independent.co.uk/news/world/europe/art-gallery-fake-painting-forgery-etienne-terrus-elne-france-a8327531.html

Waring. (2023). https://arthive.com/jacksonpollock/works/378650~Red_black_and_silver

10

Defining Meaningful Differences

Introduction

When completing a comparative analysis, it is critical to recognize that all items are unique. This is the principle of individualization, as defined by Dr. Paul L. Kirk (1902–1970), a professor of Biochemistry and Criminology at the University of California, Berkeley (Kirk, 1963). Kirk stated that "No two objects in nature are identical," which holds true for all items which may be compared. Even two items that have a common source will have some differences, possibly due to natural variations when being formed, tolerances during the manufacturing process, and/or the formation of artifacts pre- or post-production. A scientist must look at the totality of the features to determine what are meaningful versus insignificant differences when comparing two or more objects. Knowledge of the history of the material, including how it was made or created, is necessary for defining meaningful differences and distinguishing them from meaningless artifacts. Defining meaningful differences is a vital part of problem-solving, and must be considered when identifying the problem or question, making relevant observations, and forming reliable conclusions.

When making observations, spotting the insignificant differences is important so as not to waste resources or time on irrelevant analyses. This happens regularly when conducting infrared spectroscopy or microspectroscopy, where peaks due to carbon dioxide vapor ($\sim$2400–2200 cm^{-1}) are ignored during manual spectral analysis and cut from spectral library searching. It is only because the source of the peaks (i.e., carbon dioxide) is known that they can be excluded from subsequent interpretation. Similarly, when using an attenuated total reflection (ATR) objective or accessory, there may be peaks due to the uncompensated diamond ($\sim$2100–1900 cm^{-1}) which should likewise be disregarded. Regrettably, these authors have read manuscripts, both published and in peer review, where ill-informed scientists made incorrect conclusions based on these peaks which were ultimately immaterial to the problem being investigated.

Differentiating meaningful from insignificant differences is important for all light and electron microscopical analyses. The prepared mind (Chapter 6) equips a microscopist with the tools to make this determination. For example, an experienced forensic hair microscopist is able to observe numerous microscopical features of a hair (e.g., hair diameter, the

Solving Problems with Microscopy: Real-life Examples in Forensic, Life and Chemical Sciences, First Edition.
Edited by John A. Reffner and Brooke W. Kammrath.
© 2024 John Wiley & Sons Ltd. Published 2024 by John Wiley & Sons Ltd.

cuticle scale pattern, pigment granules and cortical fusi in the cortex, medullary pattern, etc.) using a comparison brightfield microscope and use these features to determine if an unknown hair could have a common source with a known individual. No two hairs are exactly the same, so the microscopist must recognize the meaningful characteristics for association or exclusion. Similarly, two bullets fired from the same firearm will not have exactly the same microscopic markings upon recovery. These differences are due to additional microscopic striations and impressions imparted to a bullet pre- and post-firing, and also due to changes to the barrel's surface with each firing. Despite this, there are microscopic surface characteristics of the gun barrel which persist and impart reproducible striations on a bullet as it travels through the barrel. It is the role of the forensic examiner to recognize the significant striations imparted to each bullet from the barrel in order to make a source identification or exclusion. In the Buttonier case (Chapter 2), a scanning electron microscope (SEM) was used to image the two halves of the broken plastic T-bar to make a fracture match, which proved that the two pieces were once connected. SEM results needed to be interpreted by a microscopist with knowledge of both how the images were created and how plastic fractures in order to scientifically evaluate the surface features imparted to the two halves of the T-bar. An important aspect of this was understanding the meaningful and insignificant differences in an SEM image of a fractured plastic surface. In this case, 18 meaningful similarities were used to make the physical match determination and insignificant differences were identified and scientifically explained.

The cases detailed in this chapter demonstrate how distinguishing meaningful from insignificant differences is critical for solving problems with microscopy.

Reference

Kirk, P. L. (1963). The ontogeny of criminalistics. *The Journal of Criminal Law, Criminology, and Police Science, 54*(2), 235–238.

10.1   The Yellow Rope

John A. Reffner, Ph.D.[1] and Brooke W. Kammrath, Ph.D.[2,3]

[1] John Jay College of Criminal Justice, New York, NY, USA
[2] University of New Haven, West Haven, CT, USA
[3] Henry C. Lee Institute of Forensic Science, West Haven, CT, USA

The body of a woman was found murdered in the basement of her Massachusetts residence. She was killed by strangulation, and the murder weapon, a yellow rope, was recovered next to the body (Figure 10.1). A boyfriend was quickly identified by detectives as a person of interest. Later, his home and property, which included a boat, were searched, and a similar appearing yellow rope was recovered from the boat and seized for comparison with the murder weapon. Due to the fact that this was a high-profile murder investigation, the physical evidence was sent to the FBI for comparison.

In the initial report from the FBI, the forensic scientist focused on the color and construction characteristics of the rope (e.g., twist, number of strands, fiber dimensions). The examiner concluded that the evidence rope and the known rope from the suspect could have had a common source due to having the same yellow color and construction. He also identified the rope as being a synthetic polyolefin, specifically polyethylene, without a complete analysis (i.e., no optical properties were measured, such as refractive index or birefringence). This physical evidence was included with other information, and used to support the charge of murder.

The defense attorneys hired a consulting criminalist with an expertise in fiber identification to examine the rope evidence. At first, a macroscopic and stereomicroscopic examination of the ropes was performed and it appeared that the association was correct. To further examine the ropes, an infrared microspectral analysis was completed in order to compare the fibers' chemistries (Figure 10.2). If the fibers came from a common source,

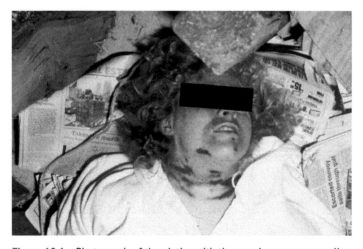

Figure 10.1   Photograph of the victim with the murder weapon, a yellow rope.

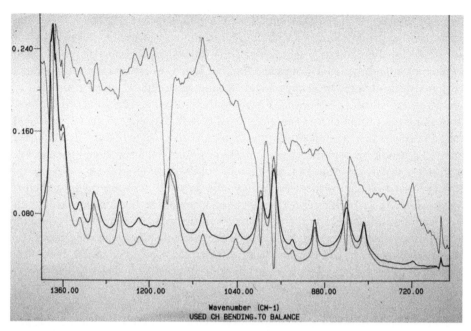

Figure 10.2 Infrared spectra of fibers from the known and questioned yellow ropes (bold black and red traces), and the result of the spectral subtraction (the top trace) on a different Absorbance scale (y-axis).

then they would have identical chemistries. An excellent way to compare two spectra is to perform a spectral subtraction to look at the differences. If the chemistries are the same, the result of the spectral subtraction would be a straight line indicating there is zero difference. If there is a uniform difference, such as would be caused by variation in thickness of the two samples, it would still result in a straight line but with a non-zero absorbance. Two important points were revealed from the infrared microspectral analysis.

First, the infrared analysis revealed that the fibers were both misidentified as polyethylene when they were in fact polypropylene. Polyethylene and polypropylene are both melt-spun olefin fibers, with similar physical and microscopic morphologies. Because both of these olefin fibers float in water, they are both commonly used to make ropes for docking lines and life-saving floatation devices. Polyethylene is more frequently encountered than polypropylene, so this assumption by the FBI examiner is understandable; however, the scientific method requires validation before making a conclusion. Polarized light microscopy could have confirmed the fiber-type classification based on measurements of the fibers' optical properties. However, this mistaken classification of both fibers did not change the ultimate conclusion of the known and questioned ropes sharing a common source. The misidentification was brought to the attention of the FBI examiner, who readily issued a revised report regarding the fiber types, but not on the association.

More interesting was the results of the infrared spectral subtraction (Figure 10.2). Several differences in the resulting spectra were observed, and required further spectral interpretation. In particular, an explanation was needed to understand a potentially meaningful difference in the C–H stretching bands at 997 and 972 cm^{-1}. When discussing these spectral differences with Norman Colthup, a world-renowned expert in infrared spectral interpretation who created the first tables showing the relationship between molecular functional groups and infrared bands, an explanation was provided. Polypropylene is produced in batches, and each batch has a slightly different distribution of stereoregular isomers. Stereoregular refers to the ordered arrangement of functional groups along the carbon chain. There are three potential arrangements (Figure 10.3). Isotactic indicates that the functional groups are all on one side of the polymer chain, syndiotactic denotes a periodic reversal of the position of the functional groups, and atactic refers to having a random orientation. Polymer tacticity has an effect on the ability to form crystalline structures, with atactic polymers resisting the formation of crystals. Although small differences in the ratio of isomers between batches would not have a significant difference in the performance of the ropes, these differences are detectible in their infrared spectra. In this case, the meaningful spectral differences correlated to differences in tacticity, and were sufficient to conclude that the fibers were composed of polymers from different batches. Thus, it was highly unlikely that the two ropes had a common source.

Despite the evidence that the rope used to commit the murder was not the same as that found on the defendant's boat, the defendant was convicted. Physical evidence is only part of the adjudication. In this case, the conclusion of the rope examination indicated that the true source of the rope was never found. This was important, but only one piece of the puzzle. Had the questioned and known ropes had a common origin, the physical evidence in the case against the defendant would have been stronger. However, failure to have a common source did not exclude the possibility that the defendant committed the crime. Further, presentation of this evidence at trial demonstrated a rigorous defense by the attorneys. Forensic scientists must remember that good science cannot be measured by a jury's verdict.

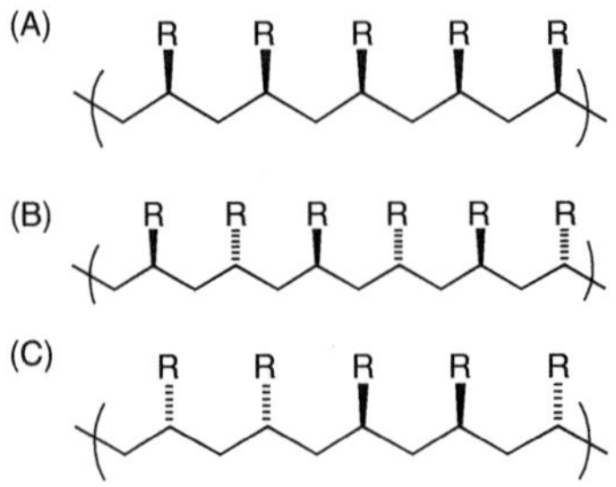

Figure 10.3 Ball-and-stick diagrams representing the orientation of an isotactic (A), and syndiotactic (B), and atactic (C) polymer chain.

10.2   Lightning Strike

John A. Reffner, Ph.D.[1] and Brooke W. Kammrath, Ph.D.[2,3]

[1] *John Jay College of Criminal Justice, New York, NY, USA*
[2] *University of New Haven, West Haven, CT, USA*
[3] *Henry C. Lee Institute of Forensic Science, West Haven, CT, USA*

On October 13, 1998, the body of 12-year-old Jorge Luis Cabrera was found at 7:15 a.m. next to a bus shelter on the intersection of Coral Way at Southwest 108th Court in Miami, Florida. The night before, the boy had an argument with his mother and ran away from his home into a lightning storm. It is believed that he sought relief from the rain in the bus shelter, where he met his unfortunate end.

Upon discovery of the body, it was apparent that the boy died from electrocution; however, the source of the energy was in question. The bus shelter was initially constructed in October 1997, and was updated to include electricity months later. This update included the addition of a transformer used to reduce the power received from a nearby light pole from 480 volts to 120 volts and wiring to illuminate an Eller Media advertising panel. Subsequent to this, overhead lighting was also installed. Humberto Codispoti, Miami-Dade's chief electrical inspector at the time, was called to the scene immediately. After an initial inspection, Codispoti quickly concluded that the culpability lied with faulty instillation of the transformer (Whitefield, 2001). This was further supported by the location of the victim, who was found with his left shin touching a metal conduit which connected the bus bench to the light pole. Third-degree burns on the victim's left shin, measured to be 2" × 2", were documented at autopsy by the medical examiner. The response to this conclusion was to inculpate Eller Media and an unlicensed electrician who worked for the company, who were both charged with manslaughter although these criminal charges were eventually dropped. A civil case did proceed. The defense developed an alternative hypothesis for the cause of the boy's death: lightning. Although there may be different clinical effects between a lightning strike and a high-voltage electrocution, there are a broad spectrum of possible injuries resulting from both. In this case, there was some ambiguity which prevented a conclusive determination. The autopsy report concluded that electrocution was the cause of death, without indicating the source of the electric energy. Thus additional evidence was required. The problem that needed to be solved was whether the boy died from electrocution from the 480-volt faulty wiring or from a lightning strike which is approximately 10,000,000–120,000,000 volts (Adams, 1987; Cutnell & Johnson, 1995; Lide, 1996–1997).

Lightning is responsible for an average of 62 deaths and 300 injuries in the United States each year, according to the National Weather Service (Infoplease, n.d.). Florida is the deadliest state with respect to lightning fatalities, despite ranking fourth for the number of lightning strikes (Pedersen, 2020). This has been attributed to the greater population density in Florida than the top three states (Texas, Oklahoma, and Kansas). From 1998 to 2020, there were 74 reported lightning deaths in Florida, as compared to only 28 in Texas, the second most deadly state, during that same time period (Infoplease, n.d.). The medical effects of a lightning strike vary greatly, with some individuals having no immediate physical signs and mild disorientation to death due to cardiac arrest and anoxic brain injury (Cooper, 2021).

A consulting criminalist was hired by the defense to provide scientific analysis of the physical evidence. Of particular value in this case was the boy's clothing, specifically a pair of shoes, a pair of black boxer briefs, and a white cotton t-shirt. One may wonder what information clothing may have in this type of case, however it is known that individuals struck with lightning will have small holes with charred edges in their clothing created by the large amount of energy. These items were brought from Florida to the private laboratory in Connecticut for forensic examination.

The white cotton t-shirt was examined first, which had been packaged in a resealable plastic bag. Upon examination of the shirt using a stereomicroscope, two different types of small holes were observed. The first were small pin holes with a dark area around it which appeared to be charred fibers. These were consistent with the characteristic holes created by a lightning strike. The second were slightly larger holes with a boundary that was a clear transparent ring of material. While observing one of these holes with the transparent ring, the head of a small termite popped up from within the folds of the shirt. It was immediately clear to the criminalist, who had prior research experience involving termites (*fortune favors the prepared mind*, Chapter 6), that the termites were the cause of the small holes with the transparent ring. Termites' diet consists of cellulose, usually sourced from wood and other plants. However, they cannot digest cellulose on their own. They rely on microorganisms such as bacteria, protozoans, and Achaea to break down the cellulose into digestible substances. This then led to the question about the source of the termites in the clothing. Termites are common pests in Florida, and they likely were the first responders to the scene of the boy's death. When the police came hours later and the body transferred to the medical examiner's office for autopsy, the termites could have remained with the body. Thus when the clothing was removed from the body, and incorrectly packaged in plastic bags (note: paper bags are recommended for evidence collection and preservation in most cases), the termites were trapped and feasted on the cotton (cellulose) clothing.

The black boxer briefs contained additional evidence that informed the scientific investigation. In the gusset or crotch area of the briefs, where sweating most commonly occurs, the fibers had been blown apart causing major damage and tearing. A monofilament nylon fiber used in the stitching thread of the underwear had a unique end morphology. When fibers are cut, a sharp edge is observed, whereas when they are torn, the edges are stretched and deformed. In rapid sheer fracture, which is a high energy event such that is caused when a bullet perforates a synthetic textile (Huemmer, 2007; Palenik et al., 2013) or when a whip is cracked and fibers are broken off, a bulbous end to the fibers develops. In these events, the high energy and short duration cause the end of the fibers to melt into short balls rather than extend along the fibers length. The end of the monofilament nylon in this case had a morphology similar to that seen with rapid sheer fracture, which indicated that a high energy short duration event caused the fiber to break. This was consistent with having been caused by a lightning strike. An electrocution from the faulty wiring in the bus shelter would not have had the energy to cause this amount and type of damage to the underwear.

The victim's tennis shoes were also examined. There was nothing remarkable about the left shoe, however the right shoe had extensive damage. The right shoe contained a large hole, roughly ¾ of an inch in diameter, on the upper portion of its right side toward the heel. The hole was heavily charred, and perforated the layers of the shoe. The damage to the shoe indicated that this was either the entry or exit location for the electrical energy. Although there was no conductive material in the construction of the shoes, given the weather at the time of the incident, the shoes were likely wet on their surface thus enabling the shoe to be the impact or exit point. A second degree burn on the victim's right foot, which included a 1.5″ × 0.75″ area of red-brown discoloration within a 3.5″ × 1.5″ area of blistered and wrinkled skin, corresponded to the damage to the right shoe.

After examining the physical evidence, the criminalist concluded that the textile damage was caused by a lightning strike rather than a high voltage electrocution. Although there are several different types of lightning strikes (Kumar et al., 2012), the damage to the clothing was consistent with what is known as the "flash effect." The flash effect occurs after there is an arc of current from a high voltage source, which results in current passing over and around the victim's body but not through it (Hettiaratchy & Dziewulski, 2004). This commonly causes both the clothes and shoes of the victim to be torn apart.

Despite the scientific findings, in June 2005, the jury in the civil case determined that the blame lied with the owner of the bus shelter, Eller Media Corp, and awarded the boy's father $4.1 million in compensatory damages and $61 million in punitive damages, for a total judgment of over $65 million. The jury concluded that the bus shelter company was negligent in the construction, installation, and maintenance of the shelter, and that these actions led to the boy's death. This sent a message to the community about the importance of electrical safety, although there were those who disagreed with the verdict. One of the defense attorney's, Ron Cabaniss, was quoted as saying "I still think that science showed the bus shelter did not cause the death of the boy, but the jury has spoken" (Tejedor, 2005). Similar to that of the case of the Yellow Rope, the jury's conclusion did not conform with the results of the scientific analysis. This again leaves the question: was justice served?

References

Adams, C. K. (1987). *Nature's Electricity* (p. 132). Tab Books.

Cooper, M. A. (2021, September 17). *Lightning Injuries Clinical Presentation: History, Physical Examination, Complications.* Medscape. Retrieved September 5, 2022, from https://emedicine.medscape.com/article/770642-clinical#b3.

Cutnell, J. D., & Johnson, K. W. (1995). *Physics* (3rd ed). Wiley. 608.

Hettiaratchy, S., & Dziewulski, P. (2004). ABC of burns: Pathophysiology and types of burns. *BMJ, 2004328*, 1427–1429. https://doi.org/10.1136/bmj.328.7453.1427

Huemmer, C. (2007). The study of rapid shear in syn- thetic fibers from ballistic impact to fabrics using po- larized light microscopy. (MS thesis). John Jay College of Criminal Justice, City University of New York.

Infoplease. (n.d.). Lightning deaths 1998-2008. Infoplease. Retrieved September 5, 2022, from https://www.infoplease.com/math-science/weather/lightning-deaths-1998-2008#:~:text=According%20to%20the%20National%20Weather%20Service%2C%20lightning%20causes,a%20building%2C%20preferably%20one%20with%20a%20lightning%20rod.

Kumar, A., Srinivas, V., & Sahu, B. P., (2012 January 3). Keraunoparalysis: What a neurosurgeon should know about it? *Journal of Craniovertebral Junction Spine*, *3*(1), 3–6. https://doi.org/10.4103/0974-8237.110116.

Lide, D. R. (1996–1997). *Handbook of chemistry and physics*. CRC Press. 14–33.

Palenik, C., Palenik, S., & Diaczuk, P. (2013). Plumbum microraptus: Definitive microscopic indicators of a bullet hole in a synthetic fabric. *Microscope*, *61*, 51–60.

Pedersen, J. M. (2020, June 30). Florida still deadliest state for lightning as storms roll into busiest time of year. Phys.org. Retrieved September 5, 2022, from https://phys.org/news/2020-06-florida-deadliest-state-lightning-storms.html

Tejedor, C. (2005, June 25). $65.1 Million verdict in boy's electrocution death against eller media corp. Colson Hicks Eidson. Retrieved September 5, 2022, from https://www.colson.com/law-firm-in-the-news/press-mentions/verdict-electrocution-death-eller-media.

Whitefield, M. (2001, February 7). Bus shelter negligence is broader, police say. Prosecutors wary of charging others. *The Miami Herald*. Accessed online from https://www.electrical-contractor.net/ESF/Articles/Bus_Shelter_Electrocution/Miami%20Herald_Bus%20shelter_2-07-01.htm.

10.3 Raman Microprobe Characterization of ZrO$_2$ Inclusions in Glass Lightguides*

Luis Soto, Ph.D.[1] and Fran Adar, Ph.D.[2,3,]*

[1] AT&T Bell Labs, Murray Hill, NJ, USA
[2] Instruments SA Inc., Metuchen, NJ, USA
[3] HORIBA Scientific, HORIBA Instruments Incorporated, Piscataway, NJ, USA
* Current affiliation

The inductively heated zirconia furnace is widely used as a heat source for fiber drawing both in the United States and in Europe. Record lengths of high-strength glass fiber have been achieved by use of the zirconia furnace (DiMarcello et al., 1984). The strength of glass fibers depends on the distribution of microscopic defects along a given length. Surface flaws related to zirconia particle inclusions can cause low-strength fiber breaks (Tariyal & Seibert, 1982). These inclusions are usually 1–10 μm in diameter and can be found embedded on the outer surface of the fiber or just lodged between the fiber and the protective polymer coating. Figure 10.4 shows a scanning electron micrograph of a fiber fracture surface with a zirconia inclusion. The characterization of these inclusions is complicated because there are two possible sources of zirconia particulate within the furnace: the yttria-stabilized zirconia susceptor and the granular zirconia insulation known as "grog," which surrounds the susceptor. In order to differentiate unambiguously between the two sources, a technique capable of positive identification of some property of both materials is needed. Raman microprobe spectroscopy is such a technique, whereas energy-dispersive X-ray analysis in the SEM is only capable of positive identification of one of the two sources. Monoclinic and tetragonal zirconia

Figure 10.4 SEM micrograph of optical fiber fracture surface showing zirconia inclusion.

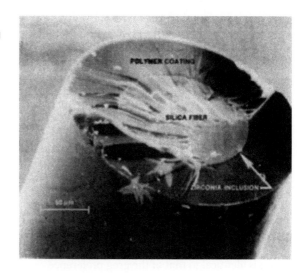

---

* *Originally published in Microbeam Analysis (San Francisco Press, Inc.) in 1984.*

have distinct and characteristic Raman spectra (Keramidas & White, 1974). The susceptor material is a mixture of the cubic and the microprobe it is possible to measure Raman spectra with high spatial resolution (Delhaye & Dhamelincourt, 1975; Etz, 1979). Clarke and Adar (1982) have shown that the technique is capable of distinguishing between the monoclinic and tetragonal polymorphs of zirconia with a spatial resolution of approximately 1 μm.

10.3.1 Experimental

In this study we examined the fracture surfaces of fibers that failed during proof-testing due to the presence of zirconia inclusions. Samples of the susceptor material and of the grog were used for reference. Powders of the reference materials were prepared by grinding with a boron carbide mortar and pestle used to avoid contamination.

Conditions, kept constant in recording all spectra, were: Excitation $\lambda = 487.987$ nm; slits, 300 μm; data points spaced by 1 cm^{-1}. The spectra of the powdered reference materials were recorded with a 50×, 0.6 NA objective; laser power = 100 mW at the sample; and counting time of 0.5 s/point. For the spectra of the inclusions a 100×, 0.9 NA objective was used; laser power = 10 mW at the sample; and counting times of 1 s/point for the spectrum shown in Figure 10.5(c) and 8 s/point for the spectrum shown in Figure 10.3C.

10.3.2 Results and Discussion

Raman spectra of the region 100–300 cm^{-1} of the susceptor, the grog, and a particle inclusion are shown in Figure 10.5. The susceptor spectrum shows two bands at 148 and 264 cm^{-1}. As mentioned previously, the susceptor material is mostly cubic zirconia with some fraction of tetragonal phase present. The Raman bands for the cubic polymorph are weak and overlap with tetragonal bands, so that the spectrum of the susceptor looks like a typical spectrum of tetragonal zirconia.

The grog spectrum shows a doublet at 181 and 192 cm^{-1} and weaker bands at 105 and 224 cm^{-1}, typical of monoclinic zirconia. A spectrum from a 2 × 4 μm zirconia inclusion on a glass lightguide also shows the 181–192 cm^{-1} doublet and the band at 224 cm^{-1}. This clearly shows that the inclusion is monoclinic zirconia, which is positive evidence that this particular inclusion came from the grog and not from the susceptor.

Laser-excited fluorescence was also measured in the susceptor material, the grog, and an inclusion. Raman and fluorescence spectra for the region 100–3000 cm^{-1} are shown in Figure 10.6. The spectrum of the susceptor material exhibits broadband fluorescence. The intensities of the Raman and fluorescence bands in the susceptor are of the same order of magnitude. The fluorescence spectra of the grog and the inclusion exhibit intense sharp lines that arise from excitations of rare-earth impurities. The strongest bands are two orders of magnitude more intense than the Raman lines. The inclusion fluorescence spectrum is identical to that of the grog preference.

The position of the fluorescence bands observed in the grog and the particle inclusion are listed in Table 10.1. All 32 lines were observed in both spectra. D'Silva and Fassel have reported X-ray excited fluorescence spectra of zirconia and uranium oxide (D'Silva

Figure 10.5 Raman spectra between 100 and 300 cm^{-1}: (a) suseptor, (b) grog, (c) inclusion.

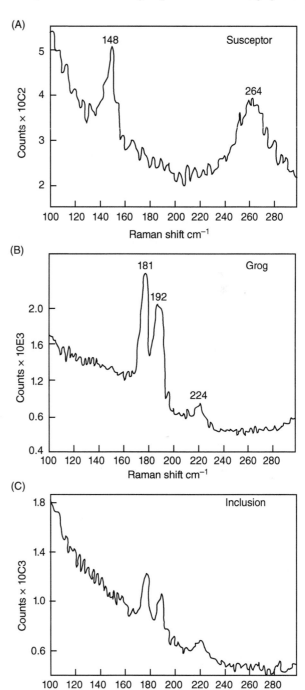

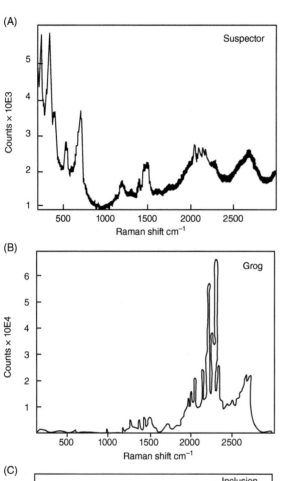

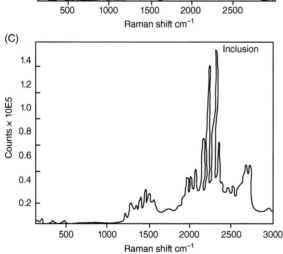

Figure 10.6 Raman and fluorescence spectra between 100 and 3000 cm^{-1}: (a) suseptor, (b) grog, (c) inclusion.

Table 10.1 Band position of fluorescence lines observed in grog and inclusion.

Shift (cm^{-1})	Wavelength (nm)
1210	518.6
1232	519.2
1265	520.1
1307	521.2
1338	522.1
1395	523.6
1453	525.2
1473	525.8
1491	526.3
1511	526.8
1563	528.3
1710	532.4
1744	533.4
1914	538.3
1940	539.0
1969	539.9
2007	541.0
2052	542.3
2091	543.4
2151	545.2
2203	546.8
2227	547.5
2258	548.4
2279	549.0
2329	550.6
2378	552.0
2449	554.2
2506	556.0
2573	558.1
2606	559.1
2664	560.9
2713	562.4

& Fassel, 1974). In both materials they observed an erbium line at 550.7 nm that corresponds to the line that we observed at 550.6 nm. This is the only line that we have been able to identify precisely. Burke and Wood have reported X-ray excited fluorescence of rare earth ions in yttria (Burke & Wood, 1967). There are four ions that fluoresce in the region 510–560 nm: terbium, erbium, holmium, and europium. These are probably the ions which give rise to the fluorescence bands observed. In comparing the positions of rare-earth ion fluorescence lines in two different materials one must keep in mind that there are shifts in band position caused by crystal field splitting. Porter et al. have studied laser-excited fluorescence spectra of the $7F_0 \rightarrow 5D_0$ transition of Eu^{+3} in eighteen different materials (Porter et al., 1983). The observed band positions vary by ± 2 nm; therefore, general comparisons of the spectra of rare earth ions in Y_2O_3 and ZrO_2 are valid.

10.3.3 Conclusions

We have shown that the Raman microprobe is an ideal tool for the characterization and differentiation of zirconia inclusions in glass lightguides. Both Raman and fluorescence spectra can be easily obtained on 1 μm inclusions. The yttria-stabilized zirconia susceptor material contains tetragonal phase, whereas the granular insulation material or grog is entirely made up of monoclinic zirconia. On the basis of these two criteria the source of the inclusions can be definitively determined.

References

Burke, W. E., & Wood, D. L. (1967). Rare-earth analyses by x-ray-excited optical fluorescence. *Advances in X-Ray Analysis, 11,* 204–213. https://doi.org/10.1154/S0376030800004869

Clarke, D. R., & Adar, F. (1982). Measurement of the crystallographically transformed zone produced by fracture in ceramics containing tetragonal zirconia. *Journal of the American Ceramic Society, 65*(6), 284–288. https://doi.org/10.1111/j.1151-2916.1982.tb10445.x

Delhaye, M., & Dhamelincourt, P. (1975). Raman microprobe and microscope with laser excitation. *Journal of Raman Spectroscopy, 3*(1), 33–43. https://doi.org/10.1002/jrs.1250030105

DiMarcello, F. V., Kurkjian, R., & Williams, J. C. (1984). Fiber drawing and strength properties. In T. Li (Ed.), *Advances in optical fiber communications.* Academic Press.

D'Silva, A. P., & Fassel, V. A. (1974). Direct determination of rare earth nuclear poisons in zirconia. *Analytical Chemistry, 46,* 996–999. https://doi.org/10.1021/ac60344a022

Etz, E. S. (1979). Raman microprobe analysis, principles and applications. *SEM/1979,* 67–82.

Keramidas, V. G., & White, W. B. (1974). Raman scattering study of the crystallization and phase transformations of ZrO2. *Journal of the American Ceramic Society, 57*(1), 22–24. https://doi.org/10.1111/j.1151-2916.1974.tb11355.x

Porter, L. C., Akse, J. R., Johnston, M. V., III, & Wright, J. C. (1983). Host Materials for Laser-excited Fluorescence of Lanthanide Probe Ions. *Applied Spectroscopy, 37*(4), 360–371.

Tariyal, B. K., & Seibert, J. M. (1982). Factors affecting low strength breaks: Fractographic analysis. *Applied Optics, 21*(10), 1716–1719. https://doi.org/10.1364/AO.21.001716

10.4 Whose Soot Is It Anyway?

Andrew Anthony Havics[1] and R. Christopher Spicer[2]

[1] *pH2, LLC, Avon, IN, USA*
[2] *Gallagher Bassett, St. Simons Island, GA, USA*

When it comes to impact from fires, there are always questions. These begin with ones like: Are there deposits from the fire (wildfire, structure fire, etc.)? Are they health hazards? Can they damage electronics? Have they created an aesthetical or perception issue (odor, color, etc.)? Sometimes these questions are answered and then it becomes: What is the source, or what is and isn't from a source? This is often the case for allocating liability.

For non-structural issues from a fire, smoke is the primary component of concern. The composition of smoke is dependent upon several factors including fuel type, moisture content of the fuel, combustion temperature, and weather-related or air-flow influences (Stone et al., 2019). Smoke from combustion of natural biomass in wildfires is a complex mixture of particulate matter, carbon dioxide, water vapor, carbon monoxide, hydrocarbons and other organic chemicals, nitrogen oxides, and trace minerals (Stone et al., 2019). The composition of structural fires is varied, with sources such as metals and plastics from electrical wiring, foams, wood, Polytetrafluoroethylene (PTFE) from coated products, textiles and carpet fibers (polypropylene, polyester, nylon, etc.), paints (metals, minerals, and polymers), polyvinyl chloride (PVC) (trim, molding, flooring, appliance covers and panels, etc.), etc.

In terms of wildfires and stack-based emissions, smoke can travel large distances depending on weather conditions. For instance, transport of large particulate from wildfires has been documented at 20 km (~12.5 mi; Pisaric, 2002) whereas combustion fly-ash spheres have been traced to 100 miles (~161 km; Flanders, 1999). The impact of these is generally small at these distances and significant deposition is generally within a few hundred yards for both wildfires and structural fires.

Thus, there are many things to consider in a fire-residue investigation. One is however limited in formal methods used to assess particulate fire residues. These are the RIA-IESO 6001–2011 Method (IESO/RIA, 2011) and the ASTM D6602 Method (ASTM, 2018). Each defines char and soot. In general, char is considered to consist of fragments of combusted material typically greater than one micron in size and are irregular in shape (IESO/RIA, 2011) or particulate larger than 1 μm made by incomplete combustion which may not deagglomerate or disperse by ordinary techniques, may contain material which is not black, and may contain some of the original material's cell structure, minerals, ash, cinders, and so forth (ASTM, 2018). Similarly, soot is defined by both as a submicron black powder generally produced as an unwanted by-product of combustion or pyrolysis. It consists of various quantities of carbonaceous and inorganic solids in conjunction with absorbed or occluded organic tars and resins (IESO/RIA, 2011; ASTM, 2018).

IESO notes that the physical characteristics described above allow char fragments to be identified by a trained microscopist using appropriate optical microscopy methods. The primary procedure used determines the presence of char particles and soot clusters (larger than 1 μm in size) on clear tape samples. Normally, a 1-cm by 1-cm area of a tape sample is marked and analyzed with a reflected light microscope using darkfield (DF) illumination at 200-times magnification. If there are no visible char or soot clusters greater than 1 μm in size (the approximate resolution of the optical microscope), a second procedure is used on wipe samples (not tape) to determine the presence of soot not initially visible on the tape samples. This involves teasing out the fibers from a wipe and mounting for microscopical analysis. Polarized light microscopy (PLM) is used at 400-times magnification to examine the mount for the presence of black soot agglomerates. Reflected light or dark field illumination is used to confirm that the particles on the wipe fibers are black and non-reflective. For the IESO/RIA method, it inherently assumes that the overwhelming majority of char and soot particles observed originated from the fire. They state that the method is not designed to provide identification of individual char particles nor to determine the origin of soot particles (IESO/RIA, 2011). Indeed, they state that if the results are not conclusively identified, other methods may be employed. If the soot particles are suspected to be unrelated to the fire (e.g., combustion of diesel fuel, candles, oil, etc.) or carbon black, additional techniques such as transmission electron microscopy (TEM) or gas chromatography-mass spectrometry (GC-MS) may be used. It is a presumptive method, not designed to differentiate or allocate sources.

The ASTM D6602 method further defines aciniform carbon as colloidal carbon having a morphology consisting of spheroidal primary particles (nodules) fused together in aggregates of <1 μm dimension in a shape having grape-like clusters or open branch-like structures. They also define carbon black as an engineered material, primarily composed of elemental carbon, obtained from the partial combustion or thermal decomposition of hydrocarbons, existing in the form of aggregates of aciniform morphology which are composed of spheroidal primary particles characterized by uniformity of primary particle sizes within a given aggregate and turbostratic (or quasi arranged parallel) layering within the primary particles.

For the ASTM D6602 method, the samples may be collected by clear tape or by a wipe (polyester wetted with a solvent). For tape samples, the tape lift is analyzed by PLM using both transmitted and reflected light. For wipe samples, a small square section (~1 cm) of a representative portion of the wipe is cut out, including an area of black staining if present. The wipe sub-sample is then placed on a slide in an appropriate liquid and analyzed by PLM and examined with transmitted (both plane-polarized light and with crossed polars) and reflected light. The analyst then estimates the percentage of each component from Table 10.4 and records this value. The identification of environmental particles and classification into categories by PLM follows that of Millette (Millette et al., 2007) and McCrone (McCrone et al., 1979) and based on an understanding of particle formation and morphology (Cottrell, 2004; Donnet, 1993; Fernandes et al., 2003; Havics and Boltin, 2007; Komarek et al., 1973; Pinorini et al., 1994; Siegla, 2013; Wolin et al., 2001; Yeoh et al., 2003). The particle types considered as potential darkening agents are: pollen, fungal, mold, biofilm, soil minerals, soot (which may include aciniform carbon, fine char, and carbon black), coal ash (fly ash), plant fragments, paint, insect parts, rust/metal flakes, rubber, coal/coke, and others. Often the microscopist observes the sample using different lighting conditions, that is, top, bottom, and side lighting. This inspection is performed in order to ascertain if carbon black, other black particles, and nonblack particles are present. Particles are examined by PLM at magnifications ranging from 100-times to 400-times.

For scanning electron microscopy (SEM) using the D6602 method, part of a sticky-tape sample is attached to an SEM sample mount (stub) with double-sided sticky tape or one transfers some of the particulate collected onto carbon adhesive material attached to an SEM sample mount (stub). The sample is placed in the SEM and viewed at 100-times to 500-times magnification observing several fields until a representative field is firmly in mind. A micrograph of the area is collected and then select certain particles of interest to test by SEM paired with energy dispersive X-ray (EDX) spectroscopy. It is important to note that the results obtained by this SEM-EDX (or wavelength dispersive X-ray spectroscopy (WDX)) technique cannot be considered as conclusive for identifying the presence of carbon black. Those with more familiarity in soot-char analysis will view the samples at up to 1000-times to 5000-times magnification as well [not required by the method].

For TEM using the D6602 Method, the sample is extracted into element-free chloroform or acetone by sonication. The resulting suspension is deposited onto a prepared carbon substrate attached to a 200- or 300-mesh copper grid. The grid is placed into the transmission electron microscope (TEM) and representative fields are examined. The aciniform materials are then evaluated for overall morphology. This is usually in the range of magnification between 5000 times and 100,000 times. The TEM imaging may be supplemented by EDX. This can be used as semi-quantitative measure of the sulfur content in individual components suspected to be carbon black, or to aid in comparing source soot material with suspect soot, or to help identify possible soot sources.

Based on the literature one can define fire particles as consisting of soot, char, ash, and tar. Soot is a fine carbonaceous material with aciniform structure; these are mostly less than 1 μm in size, but can and do occur as large aggregate chains (see definitions above). Char comprises mostly carbonaceous, large, irregular fragments of burned or partially burned material; these are mostly greater than 1 μm in size up to a few millimeters (see definitions above). Ash is the decarbonized (mostly inorganic) residue of cellulose or polymer material and typically comprises mineral salts, carbonates, oxides, or metal/non-combustible compounds. A special kind of ash, fly ash, is characteristically identified by microscopic spherical particles. Tar is observed as balls or resinous ellipsoids; it is usually associated with plant/tree material. Similar resinous balls can be observed from particulate produced by polymer or composite combustion.

Particles of all types can be generated in a fire. However, certain particles are more likely with regard to certain fuel source types. For instance, wildfires have lower soot content and higher char and ash content; and one observes tar balls with wildfires as well. On the other hand, natural gas, gasoline, and diesel and other gas or liquid fuels generally combust to form primarily soot. Building materials can generate a variety of particles depending on the fuel source and oxygen available. Tobacco particles, as a result of plant combustion, also generate char, tar, and soot. In addition, to more classical characteristics, soot also possess specific fractal dimensions (discussed below).

These fire particles (soot, char, ash, tar) thus have size, shape, color, chemical composition, and other optical properties that are used to aid in identification of fuel source and origin. These characteristics can be identified with light microscopy, SEM, TEM, EDX, Raman microscopy, and Fourier transform infrared (FT-IR) microscopy; *but no singular analysis type itself provides conclusive data on combustion origin*. An evaluation may be qualitative, semi-quantitative, or quantitative with varying degrees of certainty. Evaluations may include assemblage-type groups, principal component analysis (PCA)-like groupings, source-impact groupings, etc., each with different kinds and levels of functional differentiation of fire particulate.

In the case in question, other investigators concluded that a fire resulted in widespread contamination throughout all buildings of an apartment complex, to include interstitial spaces and cavities. Identification of fire residual in the analysis via light microscopy (modified ASTM D 6602) was recognized to be presumptive only, and inherently assumed that any combustion-related particulates identified were from the fire. Thus, there was no differentiation between actual fire-related combustion products as identified by the investigators and that which exists as recognized background contamination. Another investigation was undertaken to determine potential impact of combustion residue from the fire in wall cavities and interstitial spaces in the complex as an indicator of widespread contamination. This was done to either match or differentiate the soot sources between buildings by testing the null hypothesis that the two were similar. Sections of interior wall were removed at select locations where visual inspection and surface sampling were conducted as per a modified ASTM D 6602 by dabbing the surfaces to avoid smearing (after discussions with the lab analysts).

Twenty locations were selected across the five remaining buildings in the complex. Sample locations were selected to provide a representation of char/soot contamination in cavities at various distances from the origin of the fire that destroyed a sixth building. Sampling locations in the first four buildings were at or near perimeter walls which were accessed by cutting an approximate 16-inch by 16-inch square "coupon" into the gypsum board, followed by removal of the coupon and the enclosed batt fiberglass insulation to expose the underlying cavity surfaces for sampling. Samples were sent to RJ Lee Group (Monroeville, PA) for analysis by PLM and TEM-EDX protocols per ASTM 6602. The PLM did not reveal a means of practically differentiating the fire source (Zone 1) from the area adjacent the fire source (Zone 2).

A total of 35 aciniform-like structures from 16 samples were characterized by TEM-EDX. Four of the 20 samples yielded no aciniform-like structures. The TEM-EDX results are tabulated in terms of elemental composition in Table 10.2. These exclude Carbon (C) and Oxygen (O), which were present in all aciniform particles. Those labeled "1" were found to be present above background (3 standard deviation above noise) while those elements labeled "2" were present and they were the highest spectral peaks after carbon and oxygen.

The laboratory, upon request, selected aciniform-like particles in the sense that the particles consisted of clumped or aggregated smaller particles bound to each other to form a larger particle with poor Euclidian geometry. Despite being non-Euclidian, many were not very fractal in nature, suggesting they were not quality soot particles and may have origins other than combustion. In addition, soot is considered high carbonaceous, being primarily carbon (C) and oxygen (O) with traces of other elements depending on the source of combustion. Many of the particles had other elements with higher counts than carbon and oxygen.

Commonly one or more of potassium (K), sulfur (S), and phosphorous (P) trace elements are found in plant and wood-based combustion, whereas chlorine may be present from chloride containing polymers (such as PVC in piping or floor tile) or combined with sodium (Na) in salt aerosols. Zinc (Zn) may be found from aerosolization or corrosion particles or combustion of galvanized materials (ductwork, fan housings, flashing, railings, etc.). Thus, for instance, more than minor amounts of calcium (Ca) and sulfur (S) together suggest drywall or plaster aerosols.

Table 10.2 Elemental composition of aciniform particles.

Sample ID	Location	S	Ca	K	Na	Zn	Si	Al
1	Zone 1, Fire Source	2	1	1	1	1	1	1
1	Zone 1, Fire Source	2	1	1	1	1	1	1
2	Zone 1, Fire Source	2	1	1	1		1	1
2	Zone 1, Fire Source	2	1	1	1		1	1
2	Zone 1, Fire Source	2	1	1	1		1	
2	Zone 1, Fire Source	2	1	1	1		1	1
3	Zone 1, Fire Source	2	1	1	1		1	
4	Zone 1, Fire Source	2	1	1	1		1	
4	Zone 1, Fire Source	2	1	1	1		1	1
4	Zone 1, Fire Source	2	1	1	1		1	
5	Zone 1, Fire Source	1	1	1	1		2	1
6	Zone 1, Fire Source	2	2	1	1		1	
6	Zone 1, Fire Source	2	1	1	1		1	
6	Zone 1, Fire Source	2	1	1	1		1	
6	Zone 1, Fire Source	2	1	1	1		1	
6	Zone 1, Fire Source	2	1	1	1		1	
6	Zone 1, Fire Source	2	1	1	1		1	
10	Zone 1, Fire Source		1				1	
11	Zone 1, Fire Source	1	2	1				
11	Zone 1, Fire Source	1	2	1	1		1	1
11	Zone 1, Fire Source	1	2					
12	Zone 1, Fire Source		2	1			1	
12	Zone 1, Fire Source		2					
12	Zone 1, Fire Source		2					
14	Zone 1, Fire Source		2					
15	Zone 2, Adjacent Source	1	2					
15	Zone 2, Adjacent Source		2					
15	Zone 2, Adjacent Source		2					
16	Zone 2, Adjacent Source		2					
16	Zone 2, Adjacent Source		2					
16	Zone 2, Adjacent Source		2					
17	Zone 2, Adjacent Source		2					
18	Zone 2, Adjacent Source		2					
20	Zone 2, Adjacent Source		2					

Although the elemental composition showed strong differences between source and adjacent source, it was not sufficient to rely upon given the potential cost of remediating the soot. Given that one of the best strengths of microscopy is providing spatial distribution information, we believed that there was a means of using the aciniform morphology at a high resolution. Along this vein of reasoning is a spatial parameter called the fractal dimension (F).

F is a ratio of the perimeter to the length of the measuring scale. With the exception of very geometric shapes (square, rectangle, pentagon, etc.) many particles are fractal in nature – having a perimeter that increases as the size of the measuring stick decreases (Kaye, 1993; Kaye, 1981). Fractal dimensions may be calculated using several methods (Jin and Ong, 1995; Klinkenberg, 1994; Longley and Batty, 1989; Ramachandran and Reist, 1995), the most common being the use of the slope of a Richardson plot (Length v. Perimeter; Hayward et al., 1989) (see Figure 10.7 for an example). For computational purposes, a box counting method was used in our analysis (Hou et al., 1990; Li et al., 2009; Liebovitch and Toth, 1989) to generate a fractal dimension (F_d).

Aciniform particles are well known for having higher fractal dimensions than typical particles. Aciniform soot particles have been evaluated many times and have been found to have fractal dimensions that vary depending of the fuel source, composition, and combustion characteristics (Bonczyk and Hall, 1991; Chakrabarty, 2006; Chin-Hsiang et al., 2009; Colbeck and Wu, 1994; Donnet, 1993; Forrest and Witten, 1979; Gwaze et al., 2006; Hawa and Zachariah, 2007; Kaye, 1981; Katrinak et al., 1993; Kim et al., 2004; Luo et al., 2005; Mavrocordatos et al., 2002; Ouf et al., 2010; Pinorini et al., 1994; Sachdeva and Attri, 2008; Siegla, 2013; Skillas et al., 1998; Sorensen and Hageman, 2001; Wentzel et al., 2003; Wu et al., 2016; Xiong and Friedlander, 2001). Thus, fractal dimensions of soot have been used to determine the combustion source and processes, often in combination with other analytical testing. Examples of fractal dimensions and particle shapes are provided in Figure 10.8. Note the higher the raggedness (or complexity), the higher the fractal dimension.

We converted the laboratory's TEM images (Tiff format) to binary form and then performed box-counting fractal analysis to determine the F of the soot particles. The box counting method was applied to the images to determine F using Fraclac V 2.5 for Image J (Karperien, 2007). Examples of a Zone-1 (Adjacent Fire Source) and Zone-2 (Fire Source) aciniform images are provided in Figures 10.9 and 10.10. Fractal dimensions were analyzed on all 35 TEM particle images provided. Results are summarized in Table 10.3. Statistical data on arithmetic mean (average), standard deviation, and one-side 95% Upper Confidence Limits (UCL) were calculated and are presented in Table 10.4.

The fractal dimensions for these samples ranged from 1.45 to 1.79. This compares to fractal dimensions of other aerosols (Havics, 2019): a) diesel combustion (except when running rich) runs about 1.5–1.9, b) more pure fuel soot ranges from 1.7 to 2.0, c) intentionally fumed or carbon black ranges about 1.8–2.7, d) crustal particles (earth, soil, building materials, etc.) range from 1 to 1.6, e) biomass combustion seems to have a bimodal distribution of low Fs (1.1–1.3) and higher Fs (1.7–1.85), and e) ambient air samples contain a variety of particles ranging from 1 to 2.

The samples were collected more than a year after the date of the fire. It is recognized that a lowering of the fractal dimension occurs with aging (Dye et al., 2000; Ouf et al.,

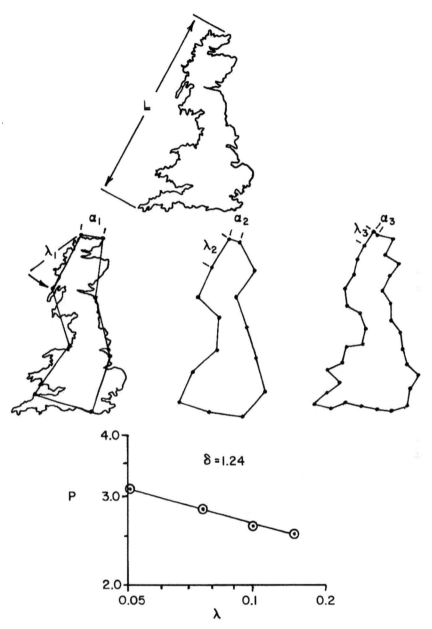

Figure 10.7 In a structured-walk procedure, a series of polygons of side λ are constructed on the perimeter using a pair of compasses. A plot of P against λ on a log-log plot yields a line with a slope |m| where δ = 1 + |m|. (*Source:* Kaye, 1989/John Wiley & Sons.)

2010). However, this is expected to be consistently the same for all samples, creating an absolute bias that does not affect a relative comparison. The fractal dimensions for Zone 1 (Fire Source) were statistically different from Zone 2 (Adjacent Fire Source) using a two-sample t-test with and without Bonferroni adjustment ($p = 0.006$ and 0.008).

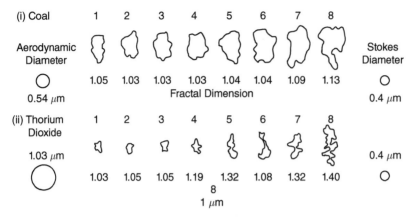

(i) Coal	1	2	3	4	5	6	7	8	
Aerodynamic Diameter									Stokes Diameter
⭕ 0.54 μm	1.05	1.03	1.03	1.03 Fractal Dimension	1.04	1.04	1.09	1.13	⭕ 0.4 μm

(ii) Thorium Dioxide 1.03 μm	1	2	3	4	5	6	7	8	
⭕	1.03	1.05	1.05	1.19	1.32	1.08	1.32	1.40	⭕ 0.4 μm

δ
1 μm

Figure 10.8 Shown above are isoaerodynamic sets of coal dust fineparticles and fumed thorium dioxide; δ is the structural boundary fractal dimension. (*Source:* Kaye, 1993/John Wiley & Sons.)

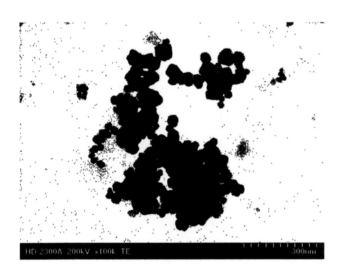

HD 2300A 200kV x100k TE   300nm

Figure 10.9 Top: Binary image of particle used for fractal dimensional analysis, Bottom: Original TEM micrograph of Particle 4 from Sample 6 from Zone 1, Fire Source.

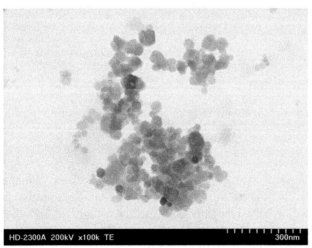

HD-2300A 200kV x100k TE   300nm

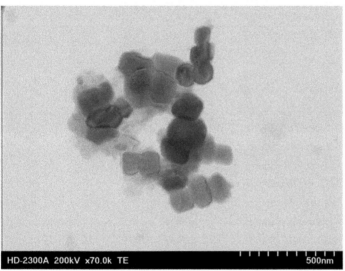

Figure 10.10 Top: Binary image of particle used for fractal dimensional analysis, Bottom: Original TEM micrograph of Particle 1 from Sample 18, from Zone 2, Adjacent Fire Source.

Additional probabilistic analysis was conducted on the F data as a confirmational step of the parametric analysis based on the mean, in evaluation of dispersal of fire-related soot/char (Spicer and Gangloff, 2000). A test hypothesis is established in which it is assumed that two zones of comparison are the same. The F data is then tested to determine if the assumption holds. Mathematically, assuming "sameness," equates to combining the data from the two zones, by which two axioms hold: 1) the median value from the combined data is the "true" central value, and 2) the frequency of values above the combined median cannot be significantly different between the two zones.

Table 10.3 Fractal dimension results by particle.

Sample ID	Location	Fractal dimension (D_f)
1	Zone 1, Adjacent Fire Source	1.756
1	Zone 1, Adjacent Fire Source	1.674
2	Zone 1, Adjacent Fire Source	1.7647
2	Zone 1, Adjacent Fire Source	1.4701
2	Zone 1, Adjacent Fire Source	1.6197
2	Zone 1, Adjacent Fire Source	1.7136
3	Zone 1, Adjacent Fire Source	1.7693
4	Zone 1, Adjacent Fire Source	1.7846
4	Zone 1, Adjacent Fire Source	1.7753
4	Zone 1, Adjacent Fire Source	1.7937
5	Zone 1, Adjacent Fire Source	1.7511
6	Zone 1, Adjacent Fire Source	1.6370
6	Zone 1, Adjacent Fire Source	1.7466
6	Zone 1, Adjacent Fire Source	1.6684
6	Zone 1, Adjacent Fire Source	1.7278
6	Zone 1, Adjacent Fire Source	1.5270
6	Zone 1, Adjacent Fire Source	1.5778
7	Zone 1, Adjacent Fire Source	1.6321
10	Zone 1, Adjacent Fire Source	1.5339
11	Zone 1, Adjacent Fire Source	1.4735
11	Zone 1, Adjacent Fire Source	1.6586
11	Zone 1, Adjacent Fire Source	1.5882
12	Zone 1, Adjacent Fire Source	1.4549
12	Zone 1, Adjacent Fire Source	1.5439
12	Zone 1, Adjacent Fire Source	1.5037
14	Zone 1, Adjacent Fire Source	1.5472
15	Zone 2, Fire Source	1.4967
15	Zone 2, Fire Source	1.5285
15	Zone 2, Fire Source	1.5614
16	Zone 2, Fire Source	1.7447
16	Zone 2, Fire Source	1.4576
16	Zone 2, Fire Source	1.4495
17	Zone 2, Fire Source	1.4574
18	Zone 2, Fire Source	1.5294
20	Zone 2, Fire Source	1.5101

Table 10.4   Statistical results on fractal dimensions.

Grouping	N	Average fractal dimension ($_{Fb}$)	Standard deviation of fractal dimension (S_{Fb})	95% upper confidence limit on fractal dimension (UCL_{Fb})
Zone 2, Fire Source	9	1.526	0.09	1.596
Zone 1, Adjacent Fire Source	26	1.642	0.111	1.687

Table 10.5   Probabilistic analysis results.

Differences in fractal dimension – probability analysis

Parameter	Median	F(A)/9	F(B)/26	P B > A
Fractal Dimension (Fb)	1.5	1	17	0.999

F(A) = frequency of detection> combined median fractal dimension in Zone 1 data
F(B) = frequency of detection > combined median fractal dimension in Zone 2 data
Median = the "middle" value of the combined fractal dimension data across both zones
PB>A = probability fractal dimension in Zone 2 is different from (greater) Zone 1

If the frequency of values is significantly different, the two zones do not exhibit similarity in particulates, and it can be inferred that combustion related particulates are from different sources. A "significant" difference between the two zones is considered to exist when the differences in detection frequency (relative to the combined median) could only occur at a random probability of 0.05 (5%) or less, which coincides with a statistically "significant" difference in the two populations being compared [Spicer, 2016]. The analysis is summarized in Table 10.5.

As can be seen, the F for the particulates in the Zone 2 data is significantly greater than for particulates in the Zone 1 area, which was directly affected by the fire. As a result, we concluded that many of the particles did not meet the classical constraints of aciniform or soot morphology, indicating they were not necessarily from combustion sources. This indicates the combustion related particulates detected in the two zones were from different sources, and it was therefore inferred there was no impact from soot generated from the fire in areas peripheral to the event's direct impact. This solved the question of origin impact succinctly.

Acknowledgments

RJ Lee of Monroeville, PA, performed the PLM and TEM-EDX analysis and provided the elemental data and all micrographs used for imaging analysis.

References

ASTM. (2018). D6602-13, Standard practice for sampling and testing of possible carbon black fugitive, emissions or other env particulate, or both. 1–22. ASTM.

Bonczyk, P. A., & Hall, R. J. (1991). Fractal properties of soot agglomerates. *Langmuir, 7*(6), 1274–1280. https://doi.org/10.1021/la00054a042

Chakrabarty, R. K. (2006). Characterization of size, morphology and fractal properties of aerosols emitted from spark ignition engines and from the combustion of wildland fuels. University of Nevada, Reno.

Chin-Hsiang, L., Whei-May, L., & Jiun-Jian, L. (2009). Morphological and semi-quantitative characteristics of diesel soot agglomerates emitted from commercial vehicles and a dynamometer. *Journal of Environmental Sciences, 21*(4), 452–457. https://doi.org/10.1016/S1001-0742(08)62291-3

Colbeck, I., & Wu, Z. (1994). Measurement of the fractal dimensions of smoke aggregates. *Journal of Physics D: Applied Physics, 27*(3), 670. https://doi.org/10.1088/0022-3727/27/3/037

Cottrell, W. H. J. (2004). *The book of fire* (2nd ed.). Mountain Press Publishing.

Donnet, J.-B. (1993). *Carbon black: Science and technology*, CRC Press.

Dye, A., Rhead, M., & Trier, C. (2000). The quantitative morphology of roadside and background urban aerosol in Plymouth, UK. *Atmospheric Environment, 34*(19), 3139–3148. https://doi.org/10.1016/S1352-2310(99)00437-9

Fernandes, M. B., Skjemstad, J. O., Johnson, B. B., Wells, J. D., & Brooks, P. (2003). Characterization of carbonaceous combustion residues. I. Morphological, elemental and spectroscopic features. *Chemosphere, 51*(8), 785–795. https://doi.org/10.1016/S0045-6535(03)00098-5

Flanders, P. J. (1999). Identifying fly ash at a distance from fossil fuel power stations. *Environmental Science & Technology, 33*(4), 528–532. https://doi.org/10.1021/es980942s

Forrest, S. R., & Witten, T. A., Jr. (1979). Long-range correlations in smoke-particle aggregates. *Journal of Physics A: Mathematical and General, 12*(5), L109. https://doi.org/10.1088/0305-4470/12/5/008

Gwaze, P., Schmid, O., Annegarn, H. J., Andreae, M. O., Huth, J., & Helas, G. (2006). Comparison of three methods of fractal analysis applied to soot aggregates from wood combustion. *Journal of Aerosol Science, 37*(7), 820–838. https://doi.org/10.1016/j.jaerosci.2005.06.007

Havics, A. A. (2019). Published data on fractal dimensions, 1–9. pH2, LLC.

Havics, A. A., & Boltin, R., (2007). *Manual for airborne & settled dust particle (ASDP) identification workshop*. McCrone Research Institute.

Hawa, T., & Zachariah, M. (2007). Development of a phenomenological scaling law for fractal aggregate sintering from molecular dynamics simulation. *Journal of Aerosol Science, 38*(8), 793–806. https://doi.org/10.1016/j.jaerosci.2007.05.008

Hayward, J., Orford, J. D., & Whalley, W. B. (1989). Three implementations of fractal analysis of particle outlines. *Computers & Geosciences, 15*(2), 199–207. https://doi.org/10.1016/0098-3004(89)90034-4

Hou, X.-J., Gilmore, R., Mindlin, G. B., & Solari, H. G. (1990). An efficient algorithm for fast O (N* ln (N)) box counting. *Physics Letters A, 151*(1–2), 43–46. https://doi.org/10.1016/0375-9601(90)90844-E

IESO/RIA. (2011). 6001, Evaluation of heating, ventilation and air conditioning (HVAC) interior surfaces to determine the presence of fire-related particulate as a result of a fire in a structure, 1–23. IESO.

Jin, X., & Ong, S. (1995). A practical method for estimating fractal dimension. *Pattern Recognition Letters*, *16*(5), 457–464. https://doi.org/10.1016/0167-8655(94)00119-N

Karperien, A. (2007).FracLac for ImageJ-FracLac advanced user's manual. http://rsb.info.nih. gov/ij/plugins/fraclac/fraclac-manual.pdf.

Katrinak, K. A., Rez, P., Perkes, P. R., & Buseck, P. R. (1993). Fractal geometry of carbonaceous aggregates from an urban aerosol. *Environmental Science & Technology*, *27*(3), 539–547. https://doi.org/10.1021/es00040a013

Kaye, B. H., (1981). *Direct characterization of fine particles*. John Wiley & Sons.

Kaye, B. H. (1989). *A random walk through fractal dimensions*, VCH Publishers.

Kaye, B. H. (1993). Applied fractal geometry and the fineparticle specialist. Part I: Rugged boundaries and rough surfaces. *Particle & Particle Systems Characterization*, *10*(3), 99–110. https://doi.org/10.1002/ppsc.19930100302

Kim, W., Sorensen, C., & Chakrabarti, A. (2004). Universal occurrence of soot superaggregates with a fractal dimension of 2.6 in heavily sooting laminar diffusion flames. *Langmuir*, *20*(10), 3969–3973. https://doi.org/10.1021/la036085+

Klinkenberg, B. (1994). A review of methods used to determine the fractal dimension of linear features. *Mathematical Geology*, *26*(1), 23–46. https://doi.org/10.1007/BF02065874

Komarek, E. V., Carlysle, T. C., & Komarek, B. B. (1973). Ecology of smoke particulates and charcoal residues from forest and grassland fires.

Li, J., Du, Q., & Sun, C. (2009). An improved box-counting method for image fractal dimension estimation. *Pattern Recognition*, *42*(11), 2460–2469. https://doi.org/10.1016/j. patcog.2009.03.001

Liebovitch, L. S., & Toth, T. (1989). A fast algorithm to determine fractal dimensions by box counting. *Physics Letters A*, *141*(8–9), 386–390. https://doi.org/10.1016/0375-9601(89)90854-2

Longley, P. A., & Batty, M. (1989). Fractal measurement and line generalization. *Computers & Geosciences*, *15*(2), 167–183. https://doi.org/10.1016/0098-3004(89)90032-0

Luo, C.-H., Lee, W.-M. G., Lai, Y.-C., Wen, C.-Y., & Liaw, -J.-J. (2005). Measuring the fractal dimension of diesel soot agglomerates by fractional Brownian motion processor. *Atmospheric Environment*, *39*(19), 3565–3572. https://doi.org/10.1016/j. atmosenv.2005.02.033

Mavrocordatos, D., Kaegi, R., & Schmatloch, V. (2002). Fractal analysis of wood combustion aggregates by contact mode atomic force microscopy. *Atmospheric Environment*, *36*(36–37), 5653–5660. https://doi.org/10.1016/S1352-2310(02)00702-1

McCrone, W. C., McCrone, L. B., & Delly, J. G. (1979). *Polarized light microscopy*. Ann Arbor Science Publishers.

Millette, J. R., Turner, W., Jr, Hill, W. B., Few, P., & Kyle, J. P. (2007). Microscopic investigation of outdoor "sooty" surface problems. *Environmental Forensics*, *8*(1–2), 37–51. https://doi. org/10.1080/15275920601180552

Ouf, F., Yon, J., Ausset, P., Coppalle, A., & Maillé, M. (2010). Influence of sampling and storage protocol on fractal morphology of soot studied by transmission electron microscopy. *Aerosol Science and Technology*, *44*(11), 1005–1017. https://doi.org/10.1080/02786826.2010.507228

Pinorini, M., Lennard, C., Margot, P., Dustin, I., & Furrer, P. (1994). Soot as an indicator in fire investigations: Physical and chemical analyses. *Journal of Forensic Science, 39*(4), 933–973.

Pisaric, M. F. (2002). Long-distance transport of terrestrial plant material by convection resulting from forest fires. *Journal of Paleolimnology, 28*(3), 349–354. https://doi.org/10.1023/A:1021630017078

Ramachandran, G., & Reist, P. C. (1995). Characterization of morphological changes in agglomerates subject to condensation and evaporation using multiple fractal dimensions. *Aerosol Science and Technology, 23*(3), 431–442. https://doi.org/10.1080/02786829508965326

Sachdeva, K., & Attri, A. K. (2008). Morphological characterization of carbonaceous aggregates in soot and free fall aerosol samples. *Atmospheric Environment, 42*(5), 1025–1034. https://doi.org/10.1016/j.atmosenv.2007.10.002

Siegla, D. (2013). *Particulate carbon: Formation during combustion*, Springer Science & Business Media.

Skillas, G., Künzel, S., Burtscher, H., Baltensperger, U., & Siegmann, K. (1998). High fractal-like dimension of diesel soot agglomerates. *Journal of Aerosol Science, 29*(4), 411–419. https://doi.org/10.1016/S0021-8502(97)00448-5

Sorensen, C., & Hageman, W. (2001). Two-dimensional soot. *Langmuir, 17*(18), 5431–5434. https://doi.org/10.1021/la0104065

Spicer, R. C., & Gangloff, H. J. (2000). A probability model for evaluating building contamination from an environmental event. *Journal of the Air & Waste Management Association, 50*(9), 1637–1646.

Spicer, R.C., & Gangloff, H.J. (2016) Permutation/randomization-based inference for environmental data. *Environmental monitoring and assessment,* 188(3), 147.

Stone, S.L., Sacks, J., Lahm, P., Clune, A., Radonovich, L., D'Alessandro, M., Wayland, M., Mirabelli, M. (2019). Wildfire Smoke, A Guide for Public Health Officials, Revised. Washington, DC: EPA.

Wentzel, M., Gorzawski, H., Naumann, K.-H., Saathoff, H., & Weinbruch, S. (2003). Transmission electron microscopical and aerosol dynamical characterization of soot aerosols. *Journal of Aerosol Science, 34*(10), 1347–1370. https://doi.org/10.1016/S0021-8502(03)00360-4

Wolin, S. D., Ryder, N. L., Leprince, F., Milke, J. A., Mowrer, F. W., & Torero, J. L. (2001). Measurements of smoke characteristics in HVAC ducts. *Fire Technology, 37*(4), 363–395. https://doi.org/10.1023/A:1012776916407

Wu, Z., Song, C., Lv, G., Pan, S., & Li, H. (2016). Morphology, fractal dimension, size and nanostructure of exhaust particles from a spark-ignition direct-injection engine operating at different air–fuel ratios. *Fuel, 185*, 709–717. https://doi.org/10.1016/j.fuel.2016.08.025

Xiong, C., & Friedlander, S. (2001). Morphological properties of atmospheric aerosol aggregates. *Proceedings of the National Academy of Sciences, 98*(21), 11851–11856. https://doi.org/10.1073/pnas.21137609

Yeoh, G., Yuen, R., Chueng, S., & Kwok, W. (2003). On modelling combustion, radiation and soot processes in compartment fires. *Building and Environment, 38*(6), 771–785. https://doi.org/10.1016/S0360-1323(03)00022-2

11

The Importance of Context

Introduction

Context provides the framework for all problem-solving, without which scientific investigations would be inefficient and directionless. Problems must be defined before they can be solved, and contextual information is essential for successfully identifying and resolving problems.

The word *context* has its origins in the Latin term *contextus*, meaning "a joining together," which was originally assimilated from *com* ("with, together") and *texere* ("to weave, to make") to form the term contexere meaning "to weave together" (https://www.etymonline.com/word/context). It is context which provides the background structure for a scientist to use the scientific method to weave or bring together observations and data to solve problems. Many of the cases discussed in the previous chapters of this book exemplify this use of context to weave together information in order to resolve problems. For example, the contextual information of finding blood on the red sweatshirt in the case of "The Hooded Sweatshirt" (Chapter 8) was important as it provided an orthogonal association regarding the value of this garment in the forensic investigation. This information encouraged careful diligence on behalf of the microscopist to examine all areas of the textile to see if there were similar fibers as those that made up the red thread found at the crime scene. In the current era, some may assert that this was biasing information; however, we vehemently disagree. The contextual information aided in the discovery of fibers with the same microscopical and chemical properties on the tag of the sweatshirt, but the microscopical and chemical analysis and comparisons were not influenced or biased. The data cannot be changed by context. The iterative process of the scientific method to verify all hypotheses ensures that results are founded in the science and not negatively influenced by context or other information. We are aware of the concerns about the potential for cognitive biases (such as context) to interfere with observations and the making of conclusions (Dror, 2017), and agree that there are some situations where some levels of context may not be relevant. However, we maintain that context provides necessary information for resolving complex problems efficiently. This chapter contains case studies which demonstrate the value of context when solving problems.

Reference

Dror, I. E. (2017). Human expert performance in forensic decision making: seven different sources of bias. *Australian Journal of Forensic Sciences*, 49(5), 541–547.

Solving Problems with Microscopy: Real-life Examples in Forensic, Life and Chemical Sciences, First Edition. Edited by John A. Reffner and Brooke W. Kammrath.
© 2024 John Wiley & Sons Ltd. Published 2024 by John Wiley & Sons Ltd.

11.1 GE Capital White-Powder Case

John A. Reffner, Ph.D.[1] and Brooke W. Kammrath, Ph.D.[2,3]

[1] John Jay College of Criminal Justice, New York, NY, USA
[2] University of New Haven, West Haven, CT, USA
[3] Henry C. Lee Institute of Forensic Science, West Haven, CT, USA

After the 9/11 terrorist attacks, the American populace was in a state of shock and filled with fear. Over the next several weeks, a near panic arose because five letters containing anthrax spores were sent to members of the American media (Tom Brokaw at NBC, The New York Post, and American Media, Inc.) and two Democratic Senators (Tom Daschle of South Dakota and Patrick Leahy of Vermont). The first letter was sent on September 18th, with authorities believing it originated from Princeton, New Jersey due to a city mailbox testing positive for anthrax spores. It must be noted that the stereomicroscope was an important tool used in the forensic analysis of these letters (Figure 11.1), as it is the first tool used in the examination of questioned documents. In what became collectively known as the Amerithrax attacks, the five letters ultimately infected 22 people, five of whom were killed, created numerous copycats who sent hoax white powders through the postal service, and also instilled a general fear and suspicion of all unknown white powders.

In the year following the Amerithrax attacks, 50,000 incidents of white-powder events were reported to the FBI (Bhattacharjee, 2012). Even now, hundreds of hoax white-powder events still occur every year. There are two types of hoax events: fake bioterrorist attacks and the public's fear of unknown white powders. Fake bioterrorist attacks take the form of a threat letter accompanied by a harmless white powder (e.g., artificial sweeteners, baking powder, corn starch, dipel insecticide, baby formula, etc.) being sent through the mail. The goal of these fake bioterrorist attacks is to cause fear and disrupt businesses, but not to actually cause personal harm. The second type is the result of a combination

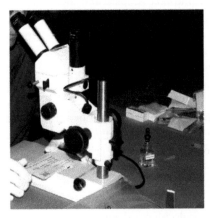 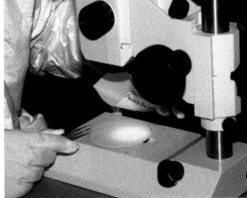

Figure 11.1 Stereomicroscopic examination of the letter and envelope sent to Senator Patrick Leahy (FBI (2010)/U.S. Department of Justice/Public Domain).

of misunderstanding, misinformation and fear. For example, an unknown white powder discovered in an office building would prompt an employee to call the authorities resulting in the building closure and a full investigation. This investigation could last several days, and the building could not open until the safety of the employees was assured by the identification of the white powder and its proper disposal.

Within a year of the Amerithrax attacks, GE Capital Corporation in Stamford, CT had a white-powder event. An employee observed an unknown white powder in the mailroom, panicked, and called the authorities. The entire building was immediately evacuated and shut down for several days for investigation. As these types of events were relatively new, there were no established hazmat procedures for first responders. The local Turner River emergency response team contacted SensIR (an analytical chemistry instrument company in Danbury, CT) and responded to the scene to investigate. A SensIR scientist, Jim Fitzpatrick, went to the GE Capital Corporation building with a TravelIR^{TM} field portable mid-infrared spectrometer. Infrared spectroscopy can be used for initial screening for anthrax since it readily detects its protein composition within a minute. Only a small amount of the white powder was necessary for analysis with this instrument, and no sample preparation was required. This was due to its pioneering attenuated total reflection (ATR) infrared technology which was made possible by a diamond internal reflection element. Within a few minutes, it was known that the unknown white powder was not anthrax, as its infrared spectrum identified it as being composed of cellulose. However, it's source was still unknown. The white powder was found near an incoming letter opening machine, thus it was hypothesized that cut envelope dust could be the source of the cellulosic white powder. Using a stereomicroscope, the paper's fibers were observed, which confirmed that the letter opening machine was the source of the unknown white powder. As a consequence of this white-powder event, panic ensued causing hundreds of thousands of dollars to be lost over what turned out to be only common paper dust.

It is interesting to note that in this case, the microscope was used to confirm the infrared identification, while the reverse is more common in most analytical protocols. This was done here because of the utility of portable infrared spectroscopy for anthrax screening at the scene. After it was determined that the material was cellulose rather than a bioterrorist agent, the microscope was used to test the context-based hypothesis which identified the source of the white powder as paper fibers from the letter opening machine.

References

Bhattacharjee, Y. (2012). The curse of the white powder: How fake bioterrorism attacks became a real problem. *Slate*. https://slate.com/technology/2012/01/white-powder-hoaxes-a-trend-in-fake-terrorism.html.

FBI. (2010, May 21). Amerithrax investigation. *FBI*. Retrieved September 5, 2022, from https://archives.fbi.gov/archives/about-us/history/famous-cases/anthrax-amerithrax.

11.2   XB-70 Valkyrie Fuel Line

John A. Reffner, Ph.D.[1] and Brooke W. Kammrath, Ph.D.[2,3]

[1] *John Jay College of Criminal Justice, New York, NY, USA*
[2] *University of New Haven, West Haven, CT, USA*
[3] *Henry C. Lee Institute of Forensic Science, West Haven, CT, USA*

The North American Aviation XB-70 Valkyrie was developed in the late 1950s/early 1960s to be the next generation supersonic long-range strategic bomber, capable of riding its own shock wave at Mach-3 speeds and altitudes of 70,000 ft (21,000 m) (B-70 Valkyrie - United States Nuclear Forces, n.d.). The high speeds achieved with the XB-70 would make it difficult to detect with radar and able to avoid interceptor aircrafts, which at the time were the only defense against bomber planes. In January of 1961, President John F. Kennedy changed the focus of the XB-70 program to research only. This was due to the high costs of production (greater than $800 million) and potential vulnerability to enemy defenses. The focus of the research was to study aerodynamics, propulsion, and the effects of high-speed and high-altitude travel on humans. Two prototypes were built for the United States Air Force Strategic Command, with the first flight occurring in 1964.

In April 1966, Walter McCrone was contacted by a representative of the United States Air Force regarding a problem with the XB-70 Valkyrie airplane. The problem was identified during a test run of the land speed of the aircraft, where the fuel stream was slowed by an obstruction in a flow-control device. Upon examination of this valve device, an unknown contaminant was observed to be blocking the flow of fuel to the engine. The plane was grounded until the solution to the problem could be found, which made this a high-priority and high-pressure situation due to contractual obligations to meet specific deadlines. Walter McCrone assigned the case to his most-trusted microscopist, John Reffner, to quickly identify the source of this problem.

A sample of the contaminant was delivered to McCrone's Chicago laboratory, along with an 8 × 10 glossy photomicrograph of the material as viewed with a stereomicroscope at ~20-times magnification. The photomicrograph showed the presence of fiberglass with green globs of a dyed coating (e.g., a paint or lacquer), which informed the scientific examination. The microscopist had seen this type of fiberglass before, as it was commonly used in the making of air filters for air conditioners and other air-flow devices. Next, the sample was examined with a stereomicroscope. The fiberglass with green globs that had been seen in the photomicrograph were observed; however, another component was also detected. This was a second type of fiberglass with a smaller diameter cross section embedded into a white material. To measure the hardness of the white material, a tungsten needle was used to probe the sample. The white material was pliable and easily deformable. This led the microscopist to conclude that the second contaminant came from a fiberglass reinforced putty. Since this came from an air filter, there were other small debris particles trapped by the two fiberglass contaminants; however, it was concluded that it was the two fiberglasses that caused the problem in the fuel line.

The results of the investigation were written into a report and faxed to the Air Force representative in order to provide them with the information they needed in a timely manner.

The next day, the representative called the microscopist and was very enthusiastic about his results. In an internal investigation that was informed by the microscopical analysis, contextual clues for solving this problem were revealed. It was discovered that during the late stages of manufacturing, workers were sent inside the chamber of the fuel tank to complete the finishing processes. One of these activities involved sealing the seams with fiberglass reinforced putty, which was the source of the white fiberglass contaminant. Further, because this work was being completed during springtime in Texas, where the temperatures can get high, an air conditioning unit with a green fiberglass filter was set up to blow cold air into chamber of the fuel tank to cool off the workers. This was the source of the fiberglass with the green globs. Since the sources of the two fiberglasses were rapidly identified, remediation efforts were able to be successfully completed. It must be recognized that the problem with the XB-70 fuel line was only able to be solved because of the microscopist's ability to microscopically detect and identify these two contaminants (see Chapter 6) combined with the contextual information from the internal investigation.

Unfortunately, a few weeks later, on June 8, 1966, news reports headlined that the XB-70 Valkyrie had crashed. When the microscopist read this on the headlines of a local newspaper, he feared that perhaps he had not actually solved the fuel line problem. When more information was provided about the crash, he learned that it was a pilot error and not a mechanical one. During a photoshoot with four other aircrafts (an F-4 Phantom, an F-5, a T-38 Talon, and an F-104 Starfighter), there was mid-air collision between the XB-70 Valkyrie and the F-104 Starfighter. This caused the F-104 Starfighter to explode in mid-air, killing its pilot immediately. The XB-70 Valkyrie went into an uncontrollable spin, and crashed to the ground killing the copilot, however the pilot was able to eject from the plane and sustained serious but nonfatal injuries. Microscopy can solve lots of problems, but not those caused by human error.

Reference

B-70 Valkyrie - United States Nuclear Forces. (n.d.). Retrieved September 5, 2022, from https://nuke.fas.org/guide/usa/bomber/b-70.htm.

11.3   Cocaine Case

John A. Reffner, Ph.D.[1] and Brooke W. Kammrath, Ph.D.[2,3]

[1] John Jay College of Criminal Justice, New York, NY, USA
[2] University of New Haven, West Haven, CT, USA
[3] Henry C. Lee Institute of Forensic Science, West Haven, CT, USA

In the early 1970s, a Hertz rental car was pulled over by a Connecticut State Trooper for a moving violation. When the Trooper approached the driver, the officer observed a suspicious package in the ash tray inside of the vehicle (Figure 11.2). Upon opening the box, the officer found a portion of a plastic straw and a small screw-top glass vial, both containing residues of a fine white powder. This item was seized for analysis to determine the identity of the unknown white powder.

At this time, the Connecticut Bureau of Investigation did not have a chemical analysis division. Although they had excellent resources for the analysis of ballistic and fingerprint evidence, there were no scientists associated with this laboratory or instrumentation capable of illicit drug identification. So, this officer brought the evidence to the Institute of Materials Science at the University of Connecticut for analysis.

The officer showed the glass vial and straw to the scientist, and asked him to identify the white powder. The scientist looked at the evidence, and said "it's cocaine." This was based on the context in which the white powder was found. Cocaine was the most common illicit drug in the 1970s, and was commonly ingested by snorting through straws. Although this was consistent with the officers' suspicion of the identity of the white powder, he wanted scientific proof.

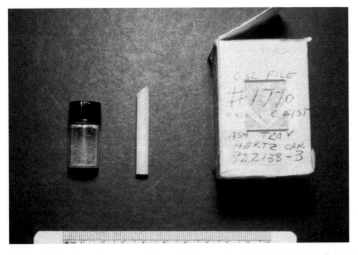

Figure 11.2   A photograph of the evidence: a small glass vial containing an unknown white powder residue, a small cut off part of a plastic straw containing a white powder residue, and the box in which they were contained.

The scientist then made an oil mount of a small sample of the unknown white powder on a microscope slide, and measured the optical properties using a polarized light microscope. This is a standard procedure that begins with choosing a specific oil from a collection of Cargille oils of various refractive indices, based on one of the principal refractive indices of the suspect material, in this case cocaine. The optical properties of common chemicals and drugs can be found in the second edition of *Optical Properties of Organic Compounds* by A.N. Winchell (1987). Cocaine hydrochloride has an orthorhombic crystal structure with $n_X = 1.522$, $n_Y = 1.594$. and $n_Z = 1.616$, thus a 1.522 oil is a good choice for the refractive index of an immersion oil. Once mounted in this oil, the optical properties of the crystal can be easily compared to the reference values. In only a few minutes, using a polarized light microscope, the hypothesis was confirmed: the sample was cocaine.

This case occurred approximately two decades prior to the formation of the Scientific Working Group for the Analysis of Seized Drugs (SWGDRUG), which was established in 1997 and cosponsored by US Drug Enforcement Administration (DEA) and the Office of National Drug Control Policy (ONDCP). It is unfortunate that these forensic drug chemists did not recognize the ability of light microscopy to obtain optical properties which can be used for drug identification. As a result, PLM was not included (and continues to be overlooked) as an approved analytical technique for drug identification.

In forensic science, there have been many meaningful discussions and conversations regarding contextual bias. Although there are valid concerns regarding biases causing false conclusions, in many cases the context can provide valuable information for achieving accurate answers to questions in a timely manner. In this case, if the optical properties of the questioned sample did not agree with the literature values for cocaine as found in Winchell's book, then it would have been concluded that the unknown materials was not cocaine. Even if it was not in this book, a known standard could be obtained, the optical properties measured and used for comparison with the unknown sample. Although microscopy is often considered subjective, measuring the optical properties of a crystal provides objective metrics for comparison and solving problems about the identity of an unknown material.

Reference

Winchell, A.N. (1987). *Optical Properties of Organic Compounds*. McCrone Research Institute, Inc.

11.4 The Preppy Murder

Peter R. De Forest[1,2] and Brooke W. Kammrath, Ph.D.[3,4]

[1] Forensic Consultants, Ardsley, NY, USA
[2] John Jay College of Criminal Justice, New York, NY, USA
[3] University of New Haven, West Haven, CT, USA
[4] Henry C. Lee Institute of Forensic Science, West Haven, CT, USA

Soil traces can provide what may prove to be surprisingly complex and valuable information, when properly interpreted. In the case to be described, the value of soil traces went unrecognized for about a year following the homicide that acquired the catchy appellation "The Preppy Murder Case" provided by the public media. It wasn't until Dr. Peter De Forest was contacted a year later and retained as a consultant by the Manhattan District Attorney's Office that the scope and value of the soil traces began to be recognized and interpreted. In the intervening year there had been some routine fragmented investigative work, and forensic laboratory work undertaken, but very little contextual information was available to the NYPD laboratory during that period. For example, information from the crime scene and/or the autopsy were not routinely shared.

No scientist had even seen or appreciated the totality of the record of traces produced during the course of the event prior to the time that one of us was retained as a private consultant to the District Attorney's Office. Much of the record had been preserved, but unfortunately it had been maintained in systemic "silos." This surviving assembled record was promptly made available to the consultant. Quite rapidly, a number of observations were integrated and hypotheses and questions for further work were developed.

On August 26th, 1986, the partially clad body of 18-year-old Jennifer Levin was found by a cyclist at approximately 6:20 a.m. in Central Park, New York, near 5th Avenue and 83rd Street, behind the Metropolitan Museum of Art. Readily visible deposits of what appeared to be soil were present on her remaining clothing and skin. The night of the 25th, Levin had gone to Dorrian's Red Hand, a popular bar for privileged young adults on the upper east side of Manhattan. Prior to entering Central Park that night, investigations showed Levin met up with friends and left Dorrian's at around 4:30 a.m. with Robert Chambers, a 19-year-old man with whom she was infatuated and had had prior sexual relations.

The crime scene investigation was fraught with missteps. The first responders drove their police car up onto the lawn where the body was found. This vehicle departed prior to the arrival of the crime scene investigators and medical examiner personnel. The presence of tire tread traces led to the incorrect initial assumption that this was a "dump job" such as the case where a prostitute was killed at another location and the body was subsequently disposed of in the park. This led to the failure of the investigators to cordon off a larger area of the crime scene to include all areas with potentially valuable traces. Some personal belongings of Ms. Levin were later recovered from an area several meters outside the initial search area. Additionally, drag marks were noted by detectives, but dismissed and thus not photographically documented by the crime scene investigators because of the false assumption that this was a "dump job." Questions as to the intention of dragging the body to a secondary location are contrary to statements made to investigators by Chambers and are outside the scope of a scientific investigation by forensic scientists, but are important to the finders of fact.

During the subsequent investigation, the police investigators went to Chambers' residence because they learned of his relationship with Levin, and suspicions about his involvement were further supported by noting fresh scratches on his face. Chambers initially denied any involvement, including claiming the scratches on his face were caused by a cat, but after being questioned further, he admitted to accidentally killing Levin during rough sex. He claimed she was manually squeezing his genitals while sitting on top of his chest, and he struck her in the neck to get her off of him. The results of the autopsy confirmed Levin's cause of death as asphyxia by strangulation.

Soil can be extraordinarily useful evidence due to its chemical and physical diversity as well as its ubiquity. Soil has been defined as "earth material that has been collected accidentally or deliberately and has some association with the matter under investigation" (Murray & Tedrow, 1992). As stated by Hans Gross (1893) in his book *Handbuch fur Untersuchungsrichter* translated as "Manual for Examining Magistrates" (which was published in English under the title *Criminal Investigation*, Gross et al., 1962), "Dirt on shoes can often tell us more about where the wearer of those shoes had last been than toilsome inquiries." Soil traces are used in forensic investigations for two purposes. First, they are used as associative evidence whereby an unknown or questioned sample is compared to a known source to determine if they share a common origin. Second, it is used for reconstruction purposes, whereby the soil traces provide information about the sequence of events or actions under investigation. In this case, the soil traces provided contributions in both of these areas.

This case demonstrates the loss of potential evidence as a result of not considering traces on the body at the initiation of an autopsy. Deposits on the victim's face and body could not be examined microscopically because the body was washed prior to the initiation of the formal autopsy. The soil traces on her face were hypothesized to be composed of a mixture of vegetable and mineral soil particles with saliva by the consulting criminalist, but this could not be confirmed due to the tendency to prioritize the autopsy over traces on the body. The presence of a criminalist with expertise in traces can be exceedingly valuable during an autopsy; unfortunately this is exceedingly rare. Configurations on the jacket (Figure 11.3) that were composed of mixtures of saliva and soil suggested that they were transferred as a result of the jacket being stuffed in her mouth as a gag during the event. This is at odds with Chamber's various statements and confessions, as it indicated intentional violence as opposed to unintentional injuries resulting from rough sex.

The body was transported to the medical examiner's office for the autopsy. Prior to the autopsy proper, the body was prepared by a diener. It was placed on the autopsy table with the remaining clothing still in place. It would have been valuable to have a criminalist scientist present to examine the body for traces and collect specimens, but this was not the case. Unfortunately, even at the time of this writing, decades later, this is not a common practice, and, in fact, it is rare.

Preliminary orientation photos were taken of the body (Figure 11.4). In addition to the soil-like traces on the clothing and body of the victim, these images revealed more subtle configurations of soiling on the face (Figure 11.5). Unfortunately, these, like all other potentially valuable deposits on the body, were washed away rather than being collected and preserved for subsequent trace microscopical analysis. Fortunately, the soil deposits on the clothing were available as reference material because the clothing had been preserved. The consulting criminalist postulated that apparent stains seen on the body and skirt in

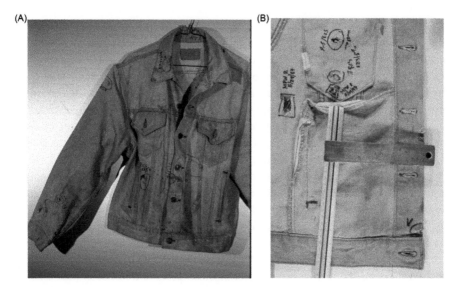

Figure 11.3 Photographs of the victim's denim jacket, showing an overall of the outer surface (A) and a midrange photograph of the inner left portion of the jacket showing deposits composed of a mixture of blood, saliva, and soil (B).

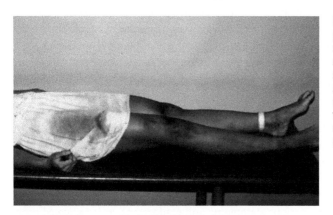

Figure 11.4 Pre-autopsy photograph of Jennifer Levin, showing an arc or boomerang-shaped region of heavier soil inside a larger area of soil deposition on the lower portion of her right skirt. Also visible next to this area is a narrow triangular-shaped void region.

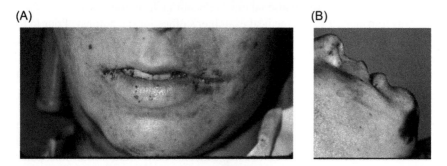

Figure 11.5 Photographs (cropped from larger images) of the victim's face as viewed from the front (A) and the right side (B), prior to autopsy showing soil deposits at the commissure of the mouth and the right cheek. The deposit on the right cheek has a distinct formation with a right-angle morphology, that was hypothesized by the consulting criminalist to potentially be a contact transfer impression from a right-angle configuration of the inside structure of the denim jacket (Figure 1B). The autopsy report indicated that the frenulum (a piece of tissue inside the upper lip) was also torn.

the initial photos were made by soil, and the geometry of these indicated she was dragged face down, head first with the upper body elevated by the assailant (Figures 11.4 and 11.6). The absence of similar deposits on the upper garments supports this theory. In addition to the deposits on the skirt, there were traces on the denim jacket (Figure 11.3). Additionally, it was postulated that some of the stains on the denim jacket which were not easily dislodged may have been made by a combination of saliva and soil. The presence of saliva was confirmed with a newly developed monoclonal antibody test specific for salivary amylase (AMY1). Further, stereomicroscopy, brightfield and polarized light microscopy revealed the stains to also contain a mixture of mineral grains and degraded vegetable material typical of soil. Combined with this biochemical and microscopical information, there was a general stellate-like pattern on the inside of the jacket that was supportive of the proposition that the jacket had been stuffed into the mouth (oral cavity) of the victim.

The photograph of the unwashed body (Figure 11.3) and her clothing (Figure 11.4) reveals a plethora of mutually supporting observations that indicate Levin was dragged face-down by her raised upper body (e.g., shoulders or arms). Her knees were heavily soiled, but the area of her lower thighs, just above the knees, was relatively unsoiled. This indicates that her upper legs were protected from contact with the soil by her skirt. It is clear the skirt was not in the position and configuration seen in the photograph (Figure

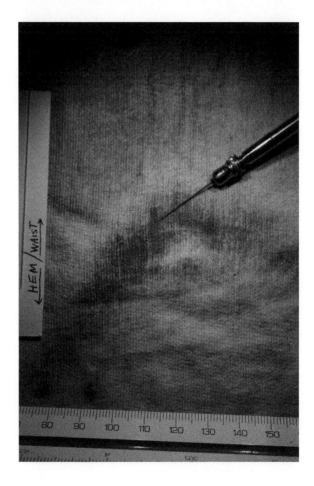

Figure 11.6 A photomicrograph of the arc or boomerang-shaped soil deposit on the skirt, viewed with a stereomicroscope.

11.3) at the time the deposits were made. The skirt was composed of jersey knit cotton fabric with an unfinished lower hem that naturally rolled up on itself. It can be seen that the rolled-up jersey material had tucked up under the skirt, resulting in the formation of a rope or cord-like contour to what became the lower edge of the skirt at the time of the dragging. The presence of the raised contour resulted in increased pressure with the ground in that area facilitating a heavier soil deposit, as seen by the boomerang-shaped stain (Figures 11.4, 11.6, and 11.7). The triangular area devoid of soil below the boomerang-shaped deposit (seen to the right of the "boomerang" in Figure 11.3) further supports

(A)

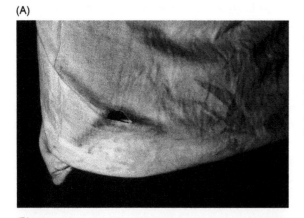

Figure 11.7 A photograph of Levin's skirt showing the location of the cutting (A) and an SEM photomicrograph at 20-times magnification of the cutting showing shingle-like downward directionality of the deformation of the yarns, which corresponded to her being dragged by the top half of her body such as her arms or hands.

(B)

this reconstruction and is a clear demonstration of the directionality of the dragging as it shows that the skirt was folded over the voided area protecting it during the dragging. It is clear that if the dragging had been foot-first, the fold would not have remained and the void would not have been created. Microscopic examination (stereomicroscopic, bright-field, and PLM) of the heavily soiled areas of the skirt and shoes, and the materials recovered from them, confirmed that they contained soil similar to that found in Central Park. This further supported that the dragging took place in Central Park rather than at some previous location.

Further support for this reconstruction was demonstrated by the directional soil deposits in the wing-tip decorative perforations on the upper anterior portion of her shoes (Figure 11.8). These deposits were heaviest in the anterior portion of each perforation, as seen in the stereomicroscope photomicrographs, which is supportive of the dragging of the victim in a face-down orientation by force applied to the upper body. Soil in the wing-tip perforations was microscopically compared and determined to be consistent with that from Central Park. It became apparent during the early part of the original scientific investigation by a NYPD forensic laboratory scientist, Nicholas Petraco, that reference soil samples were not initially collected at the crime scene because of this false initial assumption of a "dump job," and the NYPD criminalist returned to the scene shortly after to collect these necessary samples.

This case concluded with a highly publicized court trial which included nearly three months of testimony and nine days of jury deliberations. However, on March 25, 1988, with the jury at an impasse on its ninth day of deliberations, Chambers withdrew his plea of not guilty and agreed to plead guilty to manslaughter. This was part of an agreement with the district attorney, Ms. Fairstein, which included his admission in open court that he intentionally hurt Ms. Levin when he killed her. He was sentenced to 5 to 15 years in prison. Chambers, although not a model prisoner, was released in 2003 after serving his full sentence.

Ultimately, when provided with the contextual information, microscopy contributed crucial information to the resolution of this murder investigation by a skilled scientist. This case provides a prime illustration of how physical traces require context to be integrated, acquire meaning and be informative and properly understood. The scientific approach and foundation to understanding or reconstructing the circumstances surrounding the death of Jennifer Levin was not possible until a holistic approach to the investigation, aided by microscopical methods, was initiated by the generalist criminalist.

(A)

(B)

(C)

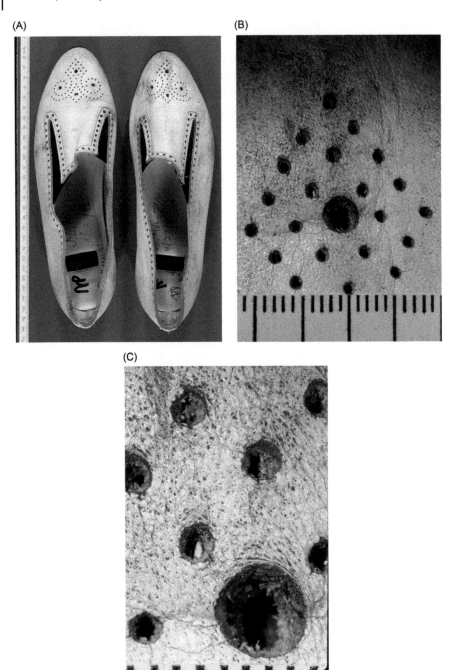

Figure 11.8 Photograph of Levin's shoes (A), and photomicrographs of the wing-tip style holes in her shoes, examined with a stereomicroscope at low and moderate magnifications (B and C). The asymmetrical deposition of soil in the holes of her shoes indicated that they had been in contact with the ground when dragged. Further, directionality of these soil deposits, in addition to the directional compression of the leather at the top of the large hole, also indicated that the victim was dragged by the top half of her body, which was consistent with the reconstruction information obtained from the soil on her skirt.

References

Gross, H. (1893). *Handbuch fur Untersuchungsrichter*. N.P.

Gross, H. (1962). *Criminal Investigation*. (Trans.) J. Adam & J. Collier Adam revised by R.L. Jackson. London.

Murray, R. C., & Tedrow, J. C. F. (1992). *Forensic Geology*. Prentice Hall.

12

Conclusion

John A. Reffner, Ph.D.[1] and Brooke W. Kammrath, Ph.D.[2,3]

[1] John Jay College of Criminal Justice, New York, NY, USA
[2] University of New Haven, West Haven, CT, USA
[3] Henry C. Lee Institute of Forensic Science, West Haven, CT, USA

12.1 Introduction

This book provides a plethora of case examples where the microscope was instrumental (pun intended) for solving a range of diverse problems. Despite the versatility of the microscope for solving problems, many scientists overlook microscopy in favor of other instrumentation. Further, there is a false perception of microscopy as being inefficient, slow, and subjective. A microscope can be simple, and elementary-to-middle-school-aged children can be taught to see magnified images. However, microscopes are sophisticated scientific instruments which require education, training, and experience to be used successfully. The prepared mind of a microscopist (Chapter 6) is able to use microscopical images as data (Chapter 3) to solve complex problems nondestructively and often in a relatively short period. Further, it is the best tool to begin a scientific investigation when used as a compass (Chapter 4) to determine subsequent analyses or courses of action. When following the scientific method (Chapter 2), observations made with the microscope reveals important features that cannot be seen otherwise. A microscope is a powerful tool for scientists to use when problem-solving.

Solving Problems with Microscopy: Real-life Examples in Forensic, Life and Chemical Sciences, First Edition.
Edited by John A. Reffner and Brooke W. Kammrath.
© 2024 John Wiley & Sons Ltd. Published 2024 by John Wiley & Sons Ltd.

12.2 Solving World Problems

Microscopy is able to address several of the world's most pressing problems. A noteworthy example of this are the initiatives being made by the inventors of the Foldscope (Figure 12.1). The Foldscope is a light microscope made from simple components, including a sheet of paper and a lens, that costs less than $1 to build. Its purpose is to be portable and durable, with no electricity requirements, while performing on par with conventional research microscopes (140-times magnification and 2-micron resolution). As part of the "frugal science" movement, the Foldscope enables communities around the world to experience the wonders and excitement of microscopy (Figure 12.2).

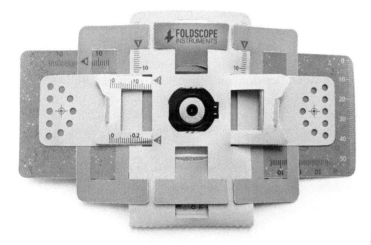

Figure 12.1   The Foldscope paper microscope. Reproduced with permission of Foldscopes Instruments, Inc.

Figure 12.2   Communities from around the world experiencing the power of microscopy using the Foldscope paper microscope. Reproduced with permission of Foldscopes Instruments, Inc.

The Foldscope, among other tools in the frugal science movement such as the Paperfuge, Octopi, Planktonscope, and Pufferfish, makes scientific tools accessible to all communities around the world (Frugal science, n.d.). The ultimate goal of the Foldscope is "to break down the price barrier between people and the curiosity and excitement of scientific exploration!" (Foldscope Instruments, Inc, n.d.). With the Foldscope, magnified images of a range of samples can be revealed, such as tiny single-celled organisms and bacteria to microscopic details in larger items like insects, fabrics, and biological tissue (Figure 12.3). The potential of this is tremendous, as no tool is more capable of sparking curiosity in young people than the microscope. When a student looks through the eyepiece of a microscope to see their cheek cells or microbes in pond water for the first time, the enthusiasm for scientific inquiry is palpable. The importance of microscopy for creating the next generation of scientists or scientifically literate citizens cannot be understated. Seeing is believing.

The Foldscope was originally developed to detect deadly blood-borne diseases such as malaria, African sleeping sickness, schistosomiasis, and Chagas in the developing world. Malaria in particular is a mosquito-borne infectious disease that is a significant world health issue. According to the World Health Organization, in 2019 there were 229 million cases of malaria with roughly 1.8% (or 409,000) fatal (World Health Organization, n.d.). Although the diagnosis of malaria can be challenging, the simple light microscope is the best tool for solving this problem. The Center for Disease Control and Prevention has the following passage on their website:

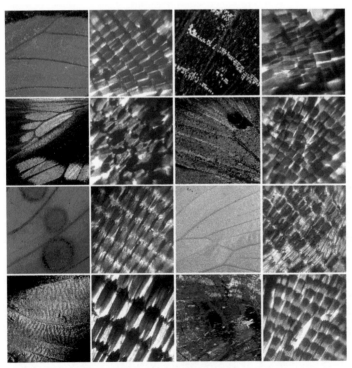

Figure 12.3 A montage of photomicrographs taken with the foldscope of various butterfly wings. (Reproduced with permission of Foldscopes Instruments, Inc).

Malaria Parasites can be identified by examining under the microscope a drop of the patient's blood, spread out as a "blood smear" on a microscope slide. Prior to examination, the specimen is stained (most often with the Giemsa stain) to give the parasites a distinctive appearance. This technique remains the gold standard for laboratory confirmation of malaria. However, it depends on the quality of the reagents, of the microscope, and on the experience of the laboratorian. (https://www.cdc.gov/malaria/diagnosis_treatment/diagnosis.html, 2021).

Despite being able to diagnose malaria with drop of a patient's blood, a major problem in developing nations is access to microscopes. They are too expensive or bulky for practical use in these under-resourced countries, and is the reason why the foldscope shows such promise. The Foldscope can be used to easily detect the characteristic malaria parasites infecting the red blood cells of a patient (Figure 12.4). This rapid diagnosis of malaria by microscopy is essential for both establishing an individual's treatment plan and stopping the spread of this deadly disease throughout the world.

In addition to the identification of blood-borne diseases, the microscope is a critical tool in many medical fields. Pathologists examine cells that have been collected by biopsy or cytology to diagnose cancer, with the type and grade of cancer often being easily identified by a trained scientist when imaged with a brightfield microscope. Dentists and endodontists use stereomicroscopes in many of their clinical procedures, including cleaning, diagnosing cavities and fracture lines, and when treating other issues such as root canals. In addition, use of the microscope, which replaced the magnifying glass, enables surgeons to perform procedures on the smallest scale possible that provides for fewer post-surgery complications and faster recovery times for patients. Microscopy is also critical in pharmaceutical research and development, which is used in the identification and investigation of pharmaceutical materials. Further, when these applications are paired with digital imaging, a reviewable record is produced which can be saved for subsequent analysis or comparisons. Modern medicine is dependent on skilled microscopists using quality microscopes to solve a vast array of health problems.

Figure 12.4 Photomicrograph of rings of P. falciparum in a thin blood smear, indicating a positive case of Malaria (U.S. Department of Health & Human Services/https://www.cdc.gov/dpdx/malaria/index.html/lastaccessed December 12, 2022).

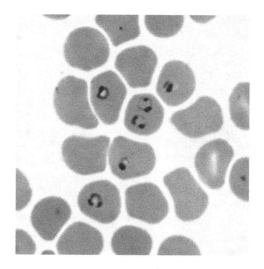

The microscope is historically one of the most important tools for solving environmental problems, such as that caused by fibrous asbestos and coal mine dust. Currently, we have a new environmental problem, microplastic pollution, for which the microscope is uniquely fit to address. Microplastics are defined as small pieces of plastic less than 5 mm in size. Trillions of tons of microplastics are estimated to be polluting our oceans, fresh water, and land. One issue is that plastics are resistant to degradation, so the accumulation of microplastics in our environment is only getting bigger. Further, due to their ubiquity, all animals, including humans, are ingesting microplastic particles through both our food and water supplies. At this time, neither the environmental effects of microplastics nor their nanotoxicity to humans or animals is fully understood. Modern research employs a range of microscopes to investigate the depth of the microplastics problem. These include stereomicroscopy for sample isolation, preparation, and initial characterization as well as infrared and/or Raman microspectroscopy for polymer identification. Without a doubt, microscopy will continue to be an essential tool for the understanding and eventual development of a solution to the microplastics problem.

12.3   Lifelong Learning

Using the microscope is a catalyst that stimulates lifelong learning. Learning does not end when a person graduates from college. People learn throughout their lives as they have new experiences. Microscopists learn throughout their microscopical investigations. This includes learning new techniques, approaches, and tips of the trade (i.e., novel sample preparation methods) in addition to the subject being explored. When education is self-initiated, this voluntary pursuit enhances our understanding of more than just the specific subject matter. All learning expands our knowledge and perception of the world. The use of the microscope inspires curiosity, boosts confidence, promotes interdisciplinary learning, and is a lot of fun.

Microscopy is seemingly easy to learn, but you continue to learn more and improve as you use it. Microscopists never stop learning. A novice microscopist often begins with "simple" microscopic methods such as stereomicroscopy and brightfield microscopy, and will seek out different ones, such as more complex contrast techniques (i.e., darkfield, phase contrast, polarized light, etc.) in order to further investigate a sample. The addition of a spectrometer or hot stage to a microscope further expands the capabilities of microscopy to extract the maximum amount of information for answering questions and solving problems, as demonstrated in several case examples. Additionally, the exploration of novel samples provides new occasions for learning the history and properties of materials (such as in the art conservation cases of "15th-century block books at The Morgan Library & Museum: the role of microscopy in unraveling complex ink formulations" by Pozzi and Basso in Chapter 4 and "Why does Guercino's *Samson Captured by the Philistines* have a grainy surface texture in some paint passages?" by Centeno in Chapter 7). The case examples in this book show how microscopy provides unlimited opportunities for lifelong learning.

12.4   Continued Evolution of Microscopy and Photonics

Technological advancements continue to improve the ability to create, record, and interpret magnified images. The history of microscopy is detailed in Chapter 1, and new technologies in microscopy and photonics continue to transform the way we observe, understand, and interact with the micro-world. Many of these new developments are due to advancements in computing power. Use of the computer has enhanced all aspects of microscopical analysis, from motorized stages and image formation to advanced image analysis and the ability to store large amounts of data. With the aid of a computer, a microscopist can interrogate a specimen in ways that could never have been done in the past. Two recent examples of these evolutionary technologies were recognized with Nobel prizes in Chemistry: the 2014 award to Eric Betzig (Janelia Research Campus, Howard Hughes Medical Institute, Ashburn, VA, USA), William Moerner (Stanford University, Stanford, CA, USA), and Stefan Hell (Max Planck Institute for Biophysical Chemistry, Göttingen, Germany, German Cancer Research Center, Heidelberg, Germany) "for the development of super-resolved fluorescence microscopy" and the 2017 award to Jacques Dubochet (University of Lausanne, Switzerland), Joachim Frank (Columbia University, New York, USA), and Richard Henderson (MRC Laboratory of Molecular Biology, Cambridge, UK) "for developing cryo-electron microscopy for the high-resolution structure determination of biomolecules in solution" (NobelPrize.org, 2014, 2017). What were once believed to be the limitations of microscopical analysis, including the abbe diffraction limit, have been circumvented. Super-resolved fluorescence microscopy, cryo-electron microscopy, and other new developments have made it possible to observe the biomolecular processes within living cells and other matrices, thus providing information that can be used to solve a range of problems including those involving cancer formation and novel drug therapies. These new microscopical technologies have made the impossible possible.

There continues to be tremendous growth in the coupling of spectroscopic methods with microscopy. In the decades since the first publications on infrared microspectroscopy in 1949 (Barer, 1949; Gore, 1949), advancements in quantum cascade lasers, scanning microscopy, and computing power have revolutionized the ability to obtain molecular information from a sample. Modern infrared and Raman microscopes (or microspectrophotometers) provide resolution on the micron-scale, Optical photothermal infrared (O-PTIR) affords sub-micron resolutions while atomic force microscopy paired with either infrared or Raman spectrometers (AFM-IR and AFM-Raman, respectively) enables nanometer resolutions (Banas et al., 2021; Reffner, 2018). These tools all have their advantages and limitations, with each having a role in scientific explorations of matter. One notable example of this is a paper by Banas, et al. in Communications Chemistry (Banas et al., 2021) which compared the three IR spectroscopic methods (IR microscopy, O-PTIR, and AFM-IR) for the characterization of red blood cells infected with malaria (i.e., plasmodium falciparum-infected human erythrocytes). The authors state "all techniques used in this work can provide quantitative distribution of bio-molecules such as DNA, carbohydrates, proteins, lipids and in the case of P. falciparum, hemozoin," (Banas et al., 2021) but the differences in spatial resolution affected the specific information obtained. The sub-micron resolution provided by O-PTIR is needed to reveal spectroscopic differences of the local biochemical variability within single cells, thus enabling the differentiation of healthy vs. infected red blood cells. Further, the nanometer resolution of AFM-IR is needed for exploring the chemical and viscoelastic properties of sub-cellular structures. It was concluded that both O-PTIR and AFM-IR "open new avenues for non-invasive monitoring of biochemical processes occurring within single cells" (Banas et al., 2021). Prior to these advancements, the limitations in spatial resolution was a barrier to the adoption of infrared spectroscopic investigations of biological samples. The capabilities of O-PTIR and AFM-IR have already provided new knowledge in a variety of scientific fields, notably the biological sciences, and these authors look forward to the many advancements in our understanding and application of sciences that will undoubtedly continue to emerge.

12.5 Final Thoughts

Microscopy has enriched our lives, both professionally and personally. Throughout our careers, the microscope has proven to be the most useful tool for solving problems. This fact was the impetus for writing this book. The world-wide community of microscopists also recognize this as fact. It is our aim with these pages to enlighten others about the value of the microscope for problem-solving through the use of case studies written by ourselves and other esteemed scientists.

At home, we have used the microscope to teach our own children about science and nature, and most importantly inspire their creativity and curiosity. Some of our favorite anecdotes include:

- When John R. Reffner (John A. Reffner's son) was in elementary school (4th grade), a study of leaves led to his learning how to use an SEM. Not only did he earn an "A" for his project, he was hooked on the power of this incredible instrument. This undoubtedly led to his career as a polymer scientist.
- When 3-year-old Drake (John A. Reffner's grandson) accidentally gained access to an unknown tablet and "tasted" it, rapid analysis with a stereomicroscope and infrared microscope enabled its identification as acetaminophen, thus quickly easing the concerns of a deeply worried mother after the hospital was unable to provide answers or reassurance. This became known in the Reffner family as "Drake's Discovery."
- Forced homeschooling during the spring of 2020 due to the COVID-19 worldwide pandemic was exceedingly difficult for parents who were unprepared for this shift in education. Riley and Grayson Kammrath's curiosity of the world was able to continue to be fostered thanks to the microscope which greatly enriched their pre-Kindergarten and Kindergarten science curriculums. Science class with mom (Brooke Kammrath) was the highlight of every day thanks to a home collection of light microscopes, and continued well into the summer break from school and beyond.

These three anecdotes further show how using the microscope to solve problems is not limited to laboratory scientists. There are limitless types of problems that can be solved with a microscope, including how to create the next generation of scientifically literate and curious citizens. Given that "curiosity is one of the great secrets of happiness" (Bryant H. McGill, author and activist) and the microscope is a catalyst for curiosity, then it is clear that a microscopist has unlocked the doors to a lifetime of happiness.

References

Banas, A. M., Banas, K., hu, T. T., Naidu, R., Hutchinson, P. E., Agrawal, R., & Breese, M. B. (2021). Comparing infrared spectroscopic methods for the characterization of plasmodium falciparum-infected human erythrocytes. *Communications Chemistry, 4*(1), 1–12.

Barer, D. D. (1949). *Nature, 163*, 198–201. (1949).

Centers for Disease Control and Prevention. (2020, October 6). *CDC - dpdx - malaria*. Centers for Disease Control and Prevention. Retrieved September 5, 2022, from https://www.cdc.gov/dpdx/malaria/index.html.

Foldscope Instruments, Inc. (n.d.). *Foldscope paper microscopes: Portable microscopes.* Foldscope Instruments, Inc. Retrieved September 5, 2022, from https://foldscope.com

Frugal science. (n.d.). Frugal Science. Retrieved September 5, 2022, from https://www.frugalscience.org

Gore, R. C. (1949). *Science, 110,* 710–711.

NobelPrize.org. (2014). The nobel prize in chemistry 2014. Retrieved September 5, 2022, from https://www.nobelprize.org/prizes/chemistry/2014/summary

NobelPrize.org. (2017). The nobel prize in chemistry 2017. Retrieved September 5, 2022, from https://www.nobelprize.org/prizes/chemistry/2017/press-release/#:~:text=The%20Nobel%20Prize%20in%20Chemistry%202017%20is%20awarded,era.%20A%20picture%20is%20a%20key%20to%20understanding

Reffner, J. (2018). Advances in infrared microspectroscopy and mapping molecular chemical composition at submicrometer spatial resolution. *Spectroscopy, 33*(9), 12–17. https://www.spectroscopyonline.com/view/advances-infrared-microspectroscopy-and-mapping-molecular-chemical-composition-submicrometer-spatial

World Health Organization. (n.d.). *Fact sheet about malaria.* World Health Organization. Retrieved September 5, 2022, from https://www.who.int/news-room/fact-sheets/detail/malaria

Index

Solving Problems with Microscopy: Real-life Examples in Forensic, Life and Chemical Sciences, First Edition.
Edited by John A. Reffner and Brooke W. Kammrath.
© 2024 John Wiley & Sons Ltd. Published 2024 by John Wiley & Sons Ltd.